XX International Youth Scientific School on Coherent Optics and Optical Spectroscopy (COOS 2016)

Journal of Physics: Conference Series Volume 859

Kazan, Russia
18 – 20 October 2016

ISBN: 978-1-5108-4194-9
ISSN: 1742-6588

Printed from e-media with permission by:

Curran Associates, Inc.
57 Morehouse Lane
Red Hook, NY 12571

Some format issues inherent in the e-media version may also appear in this print version.

Printed by Curran Associates, Inc. (2017)

For permission requests, please contact the Institute of Physics at the address below.

Institute of Physics
Dirac House, Temple Back
Bristol BS1 6BE UK

Phone: 44 1 17 929 7481
Fax: 44 1 17 920 0979

techtracking@iop.org

Additional copies of this publication are available from:

Curran Associates, Inc.
57 Morehouse Lane
Red Hook, NY 12571 USA
Phone: 845-758-0400
Fax: 845-758-2633
Email: curran@proceedings.com
Web: www.proceedings.com

Table of contents

Volume 859

XX International Youth Scientific School "Coherent Optics and Optical Spectroscopy"

18–20 October 2016, Kazan, Russian Federation

Accepted papers received: 26 May 2017
Published online: 2 June 2017

Preface

Papers

IOP Conf. Series: Journal of Physics: Conf. Series **859** (2017) 011001 doi:10.1088/1742-6596/859/1/011001

The XX International Youth School on Coherent Optics and Optical Spectroscopy (COOS2016) was held in Kazan, Russia, from October 18 to October 20 on the Nikolai Lobachevsky Scientific Library of Kazan Federal University. The School follows the global tendency to comprehensive studies of matter properties and its interaction with electromagnetic fields. Since 1997 more than 100 famous scientists from USA, Germany, Ukraine, Belarus and Russia had plenary lectures presentations. This is the right place, where over 1000 young scientists had an opportunity to participate in hot discussions of the latest scientific news. Many young people have submitted interesting reports on photonics, quantum electronics, laser physics, quantum optics, traditional optical and laser spectroscopy, non-linear optics, material science and nanotechnology. Here we are publishing the full-size papers prepared from the most interesting lectures and reports selected by the Program Committee of the School. Plenary sessions were offered by the following invited speakers:

Ildar Gabitov, University of Arizona, USA.

- Error statistics in coherent communication lines

Andrei Naumov, Institute for Spectroscopy RAS, Troitsk, Moscow, Russia.

- Revisiting the question of the experimental realization of a nonclassical light source on the basis of single organic molecules of dyes

Gerd Hermann, University of Giessien, Germany.

- Applications of Coherent Spectroscopy

Askhat Basharov, National Research Center 'Kurchatov Institute', Moscow, Russia.

- Low-frequency emission in resonant processes
- Evolution of a two-level quantum particle in the noise classical e.-m. field within and beyond the resonant approximation

Anastas Bukharaev, Kazan E. K. Zavoisky Physical -Technical Institute, Kazan, Russia.

- Straintronics

Maxim Gladush, Institute for Spectroscopy RAS, Troitsk, Moscow, Russia.

- Fluorescent properties of single quantum emitters and their ensembles in dielectric media

Sergey Sazonov, National Research Center "Kurchatov Institute", Moscow, Russia.

- On the approximations of slowly varying envelope and slowly varying profile in nonlinear optics equations

Alexey Kalachev, Kazan E. K. Zavoisky Physical -Technical Institute, Kazan, Russia.

- Quantum memory in an isotopically pure crystals

Renat Gainutdinov, Kazan Federal University, Kazan, Russia.

- Atomic hydrogen spectroscopy and problems of modern physics applications

Vladimir Dresvyanskiy, Irkutsk branch of the Institute of Laser Physics SB RAS, Irkutsk, Russia.

- Photophysical processes in single radiation defects and their ensembles

Ilfan Bikmaev, Kazan Federal University, Kazan, Russia.

- Optical spectroscopy in astrophysics

This proceedings volume contains papers reported at the conference. The Proceedings of the XX International Youth School on Coherent Optics and Optical Spectroscopy (COOS2016) were edited by Salakhov Myakzyum Khalimullovich, Samartsev Vitalii Vladimirovich and Gainutdinov Renat Khamitovich with the assistance of Petrova Alexandra Sergeevna as copy editor.

We are grateful to Russian Foundation for Basic Research, Tatarstan Academy of Science, Kazan E.K. Zavoisky Physical-Technical Institute for financial.

Salakhov Myakzyum Khalimullovich, Editor
Samartsev Vitalii Vladimirovich, Editor
Gainutdinov Renat Khamitovich, Editor

IOP Conf. Series: Journal of Physics: Conf. Series **859** (2017) 011002 doi:10.1088/1742-6596/859/1/011002

Peer review statement

All papers published in this volume of *Journal of Physics: Conference Series* have been peer reviewed through processes administered by the proceedings Editors. Reviews were conducted by expert referees to the professional and scientific standards expected of a proceedings journal published by IOP Publishing.

Transmission and reflection of self-assembled hybrid plasmonic-photonic crystals

A A Akhmadeev[1,2], A V Koryukin[1,2], A F Valitova[1] and M Kh Salakhov[1,2]

[1] Kazan Federal University, Institute of Physics, 420008, Kremlevskaya St, 18, Kazan, Russian Federation

[2] Tatarstan Academy of Sciences, 420111, Bauman St, 20, Kazan, Russian Federation

E-mail: akoryukin@gmail.com

Abstract. We studied the optical properties of hybrid plasmonic-photonic crystals using spectral photometry. These structures consist of colloidal photonic crystal and gold layer deposited onto its surface. The samples of plasmonic-photonic crystals prepared with relatively simple method allow to observe different effects. Several peaks in the transmission spectra corresponding to the extraordinary transmission were investigated. The extraordinary transmission which depends on the angle of light incidence and polarization suggests the excitation of surface plasmon polaritons. An extraordinary reflection was also observed.

1. Introduction

Plasmon-photonic crystals (PPCs) are metal-dielectric structures the dielectric components of which are photonic crystals (PCs). Such structures allow to excite plasmonic and photonic modes [1]. Photonic crystals are materials with spatial periodicity in their dielectric function (or, equivalently, a periodical index of refraction) [2]. Such periodicity yields a strong modification of the electromagnetic field that manifests itself in new quantum electrodynamical effects [3–5]. PPCs allow to increase the field confinement of PCs while overcoming short propagation distances of surface plasmon polaritons [1]. Surface plasmons are collective oscillations in electron density at the surface of a metal [6]. Surface plasmon polaritons (SPPs) are electromagnetic excitations propagating at the interface between a dielectric and a conductor [7]. PPCs are used for easily coupled whispering gallery plasmons [8], surface-enhanced Raman spectroscopy [9, 10], coherent fluorescence emission [11], spasers [12] and Tamm plasmon lasers [13]. PPCs can be prepared from one-dimensional [13–16], two-dimensional [18–22] and three-dimensional [23–30] PCs. The metal coating can be made of silver, gold and other plasmon materials. In this article we investigated the PPC based on 3D colloidal PC with gold coating.

The main feature of these structures is the extraordinary transmission and the possibility of excitation of various types of plasmons [24]. To observe these effects it is necessary to prepare a sample of high quality with the required parameters. In our work a relatively simple method of fabrication of such structures was used. This method has minimal requirements to the resources and equipment of the laboratory. We used spectral photometry to study the optical properties of the PPC samples. According to the spectra, a conclusion about the observation of extraordinary transmission and excitation of plasmons was made.

2. Sample fabrication

PPC sample was fabricated in three steps. The first step is the synthesis of silica particles by the Stober-Fink-Bohn method [31]. At the second step opal-like PC was prepared by vertical deposition of particles under the action of capillary forces [32]. The third step was the deposition of gold on the PC surface. Scheme of the surface layer of the sample is presented in figure 1. The diameter of the silica particles was 334 ± 5 nm, thickness of the PC sample – 3.6 ± 0.2 μm, number of layers – 12. These parameters were determined by atomic force microscopy and spectrophotometry [33]. The thickness of the gold layer was about 50 nm.

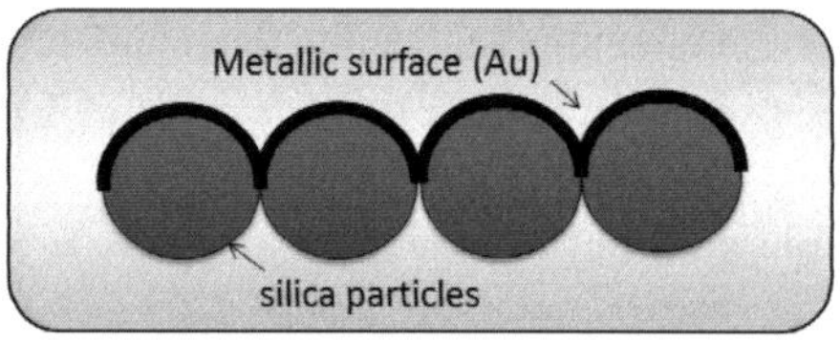

Figure 1. Scheme of the surface layer of the sample: a gold layer deposited on surface of colloidal photonic crystal.

3. Optical characterization

Spectra were recorded using UV-VIS spectrophotometer Lambda-35 (Perkin Elmer). Figure 2 shows that the transmission spectrum of the sample changed after the deposition of the gold film. For a detailed study of plasmonic modes the PPC spectrum normalized to the PC spectrum was considered. As can be seen from figure 3, this spectrum has several transmission peaks (around 350, 400 and 500 nm). They correspond to plasmon modes in different interfaces: air-metal, metal-opal. Such a spectrum evidences about the extraordinary transmission.

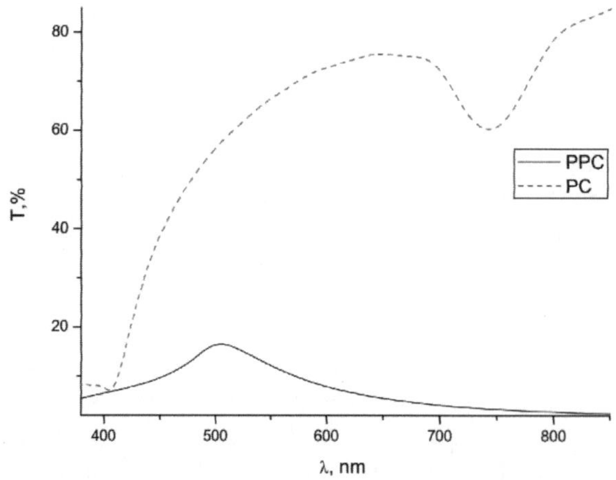

Figure 2. Transmission spectra of PPC and PC obtained at 0° of incident angle.

Figure 4 illustrates the reflection spectra showing extraordinary reflection. There is also an absorption band, which may indicate the excitation of localized plasmons.

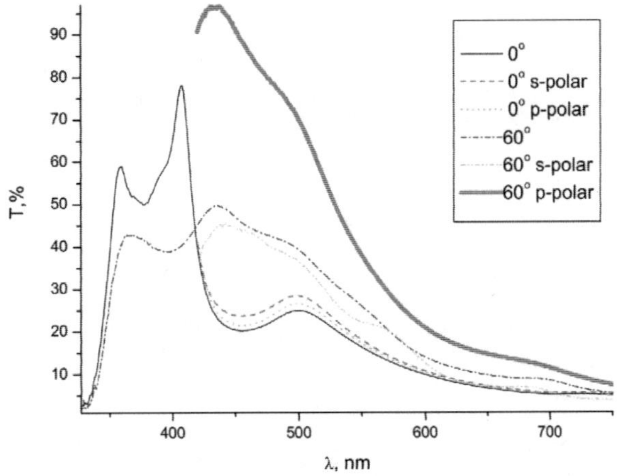

Figure 3. Transmission spectra of PPC after its normalization to transmission spectra of the PC. Spectra were taken at 0° and 60° of incident angle for various polarizations.

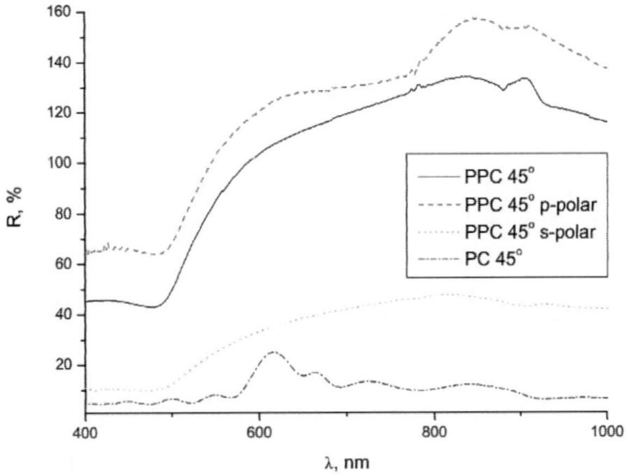

Figure 4. Reflection spectra of PPC at 45° of incident angle for various polarizations and reflection spectrum of PC.

4. Conclusion

In this article, we studied the optical properties of hybrid colloidal plasmonic-photonic crystals. The crystals were prepared with relatively simple method. Optical spectra show that, due to excitation of plasmon modes, the samples have the extraordinary transmission, which is consistent with the literature data [23–25]. The plasmonic-photonic crystals may be useful in surface-enhanced Raman spectroscopy for wavelengths over 600 nm.

Acknowledgments

We thank S. Kharintsev for the gold deposition.

References

[1] Lopez-Garcia M 2011 *Self Assembled Photonic-Plasmonic Crystals for Light control at the Nanoscale* p 185

IOP Conf. Series: Journal of Physics: Conf. Series **859** (2017) 012001 doi:10.1088/1742-6596/859/1/012001

[2] Joannopoulos J D, Johnson S G, Winn J N and Meade R D 2008 *Photonic crystals: molding the flow of light* (Princeton: Princeton University Press) p 286
[3] Gainutdinov R Kh, Khamadeev M A and Salakhov M Kh 2012 *Phys. Rev.* A **85** 053836
[4] Gainutdinov R Kh, Khamadeev M A and Salakhov M Kh 2013 *J. Phys.: Conf. Ser.* **478** 012017
[5] Akhmadeev A A, Gainutdinov R Kh, Hermann G, Khamadeev M A, Steryakov O V and Salakhov M Kh 2015 *J. Phys.: Conf. Ser.* **613** 012005
[6] Novotny L and Hecht B 2006 *Principles of nano-optics* (New York: Cambridge Univrsity Press) p 539
[7] Maier S 2007 *Plasmonics: Fundamentals and Applications* (Springer Science+Business Media LLC) p 224
[8] Cole R M, Sugawara Y, Baumberg J J, Mahajan S, Abdelsalam M and Bartlett P N 2006 *Phys. Rew. Lett.* **97** (13) 137401-1-4
[9] Oh Y-J, Park S-G, Kang M-H, Choi J-H, Nam Y and Jeong K-H 2011 *Small* **7** (2) 184-188
[10] Tsvetkov M Y, Khlebtsov B N, Khanadeev V A, Bagratashvili V N, Timashev P S, Samoylovich M I and Khlebtsov N G 2013 *Nanoscale Research Letters* **8** (250) 1-9
[11] Shi L, Yuan X, Zhang Y, Hakala T, Yin S, Han D, Xhu X, Zhang B, Liu X, Torma P, Lu W and Zi J 2014 *Laser and Photonics Reviews* **8** (5) 717-725
[12] Oulton R, Sorger V, Zentgraf T, Ma R, Gladden C, Dai L, Bartal G and Zhang X 2009 *Nature* **461** 629-632
[13] Symonds C, Lheureux G, Hugonin J P, Greffet J J, Laverdant J, Brucoli G, Lemaitre A, Senellart P and Bellessa J 2013 *Nano Lett.* **13** 31793184
[14] Kaliteevski M, Iorsh I, Brand S, Abram R A, Chamberlain J M, Kavokin A V and Shelykh I A 2007 *Phys. Rev.* B **76** 165415
[15] Sasin M E *et al* 2010 *Superlattices and Microstructures* **47** 4449
[16] Afinogenov B I, Bessonov V O, Nikulin A A and Fedyanin A A 2013 *Appl. Phys. Lett.* **103** 061112
[17] Chen Y, Zhang D, Zhu L, Wang R, Wang P, Ming H, Badugu R and Lakowicz J R 2014 *Optica* **1** (6) 407-413
[18] Lopez-Garcia M, Galisteo-Lopez J F, Blanco A, Sanchez-Marcos J, Lopez C and Garcia-Martin A 2010 *Small* **6** (16) 1757-1761
[19] Lopez-Garcia M, Galisteo-Lopez J F, Blanco A, Lopez C and Garcia-Martin A 2010 *Adv. Funct. Mater.* **20** 4338-4343
[20] Galisteo-Lopez J F, Lopez-Garcia M, Lopez C and Garcia-Martin A 2011 *Appl. Phys. Lett.* **99** 083302-1-3
[21] Galisteo-Lopez J F, Lopez-Garcia M, Blanco A and Lopez C 2012 *Colloidal Nanoplasmonics* **28** 9174-9179
[22] Liu T-l, Russel K J, Cui S and Hu E L 2014 *Optics Express* **22** (7) 8219-8225
[23] Wang Z, Ye Y, Zhang Y and Zhang J 2009 *Appl. Phys.* A **97** 225-228
[24] Ding B, Pemble M E, Korovin A V, Peschel U and Romanov S G 2010 *Phys. Rev.* B **82** 035119
[25] Frederich H, Wen F, Laverdant J, Coolen L, Schwob C and Maitre A 2011 *Optics Express* **19** (24) 24424-24433
[26] Ding B, Bardosova M, Pemble M E, Korovin A V, Peschel U and Romanov S G 2011 *Adv. Funct. Mater.* **21** 4182-4192
[27] Romanov S G, Korovin A V, Regensburger A and Peschel U 2011 *Adv. Mater.* **23** 2515-2533
[28] Romanova A S, Korovin A V and Romanov S G 2013 *Physics of the Solid State* **55** (8) 1725-1732
[29] Lin T, Lin J, Guo J and Kan H 2015 *Adv. Funct. Mater.* **3** 1470-1475
[30] Korovin A V and Romanov S G 2016 *Phys. Rev.* B **93** 115440
[31] Stober W, Fink A and Bohn E 1968 *J. Colloid Interface Sci.* **26** 62-69
[32] Vlasov Y A, Bo X-Z, Sturm J C and Norris D J 2001 *Nature* **414** 289–293
[33] Koryukin V M, Akhmadeev A A and Salakhov M Kh 2013 *J. Phys.: Conf. Ser.* **478** 012013

IOP Conf. Series: Journal of Physics: Conf. Series **859** (2017) 012002 doi:10.1088/1742-6596/859/1/012002

The influence of collisions on the temporary shape of stimulated echo hologram in gas

E N Akhmedshina[1], L A Nefediev[1], G I Garnaeva[1] and E D Shigapova[1]

Department of Educational Technologies in Physics, Kazan Federal University, Kazan, 420008, Russia

Email: ekanika8@gmail.com, nefediev@yandex.ru, guzka-1@yandex.ru, elvshi@mail.ru

Abstract. In this paper, we investigate the influence of collisions with the change of particle velocity direction in a gas on the reproduction of the temporary shape of the object laser pulse in the stimulated echo hologram response. Due to such collisions, the frequency shifts of the radiation of atoms in the gas randomly vary (spectral diffusion within the heterogeneously broadened line). It is shown, that such diffusion leads to the not correlated heterogeneous broadening in the gas at the different time intervals and the partial loss of system phase memory, which results in a partial loss of retrieved information encoded in the temporal form of the object laser pulse.

1. Introduction

In gas media, the formation of stimulated echo-holograms (SEH) is qualitatively different from their analogs in a solid object. The differences are associated with the movement of "working" particles, which leads to the need to take into account the collisions and random reorientations of particles in the formation of optical coherent responses of the medium. In a solid object, each "working" particle has its own transition frequency (heterogeneous broadening of the resonance line). During the formation of photon echo signal in a solid object, at each point within the sample is a phasing of the induced dipole moments during the echo response appearance and this phasing does not depend on the excitation direction. In gases, however, velocities of the individual particles are vector quantities, so each point within the sample is associated with its own direction in the space (the directed nature of the Doppler effect).Therefore, the amplitude-phase information contained in the object wave turns out to be "encoded" in the streams of particles moving with certain velocities. In addition, the phasing of individual dipole moments in the gas at the time of echo formation does not necessarily occur at each point of the sample, but phasing of the different velocity "packs" of the particles moving with various velocities [1-6]. Elastic collisions lead to random changes in the angle between the direction of the observation and the velocity of the particle, which in turn leads to random shifts in the emission frequency of each particle (spectral diffusion). This can have an effect on the formation of the SEH, even in case of the parallel wave vectors of the exciting laser pulses. Therefore, it is in the interest to consider the effect of the spectral diffusion on the temporal form of the SEH response [2-6].

2. The basic equations

The description of the formation of the echo-holograms in gases requires simultaneous consideration of the Doppler shifts in the frequencies of the particles emission, changes in their position and the orientation in space, collisions of particles, and also degeneracy of their energy levels over some quantum number. In case of the impact mechanism of the broadening of the resonance transition line, the equation for the density matrix in the laboratory coordinate system has the form [2]:

$$\left(\frac{\partial}{\partial t}+\frac{\partial\Delta\vec{r}}{\partial t}\nabla\right)\rho_{\gamma\alpha}^{J_1J_2}\left(1-\frac{\vec{n}}{c}\frac{\partial\Delta\vec{r}}{\partial t}\right)^{-1}=\frac{i}{\hbar}\left(H_{J_1}^0-H_{J_2}^0\right)\rho_{\gamma\alpha}^{J_1J_2}+$$
$$+\frac{i}{\hbar}\sum_{\beta}\left(V_{\gamma\beta}^{J_1J}\rho_{\beta\alpha}^{JJ_2}-V_{\beta\alpha}^{JJ_2}\rho_{\gamma\beta}^{J_1J}\right)-N\left\langle \upsilon\left(\sigma'_{\gamma\alpha}-i\sigma''_{\gamma\alpha}\right)\right\rangle\rho_{\gamma\alpha}^{J_1J_2} \tag{1}$$

where J_i is the total moment of i-th level, indexes α, β, γ denote sublevels degenerate levels, $\Delta\vec{r}$ is the displacement of the particle due to its movement on the considered time interval, $\vec{n}$ is the direction of propagation of the laser excitation pulse, $V_{\beta\beta'}^{JJ'}$ are the matrix elements of the interaction operator of the atom with electromagnetic radiation, H^0 is Hamiltonian of the unperturbed atom, N is the concentration of the perturbing particles, υ is the relative velocity of the atom and the perturbing particle, σ 'and σ" are cross-sections for the broadening and shifting of the corresponding spectral line, c is the velocity of propagation of the excitation laser pulse in the medium.

The spectral heterogeneity in the gas is due to the molecular velocity spread and the dependence of the frequency shift Δ_j of the j-th molecule on the velocity (the Doppler effect). In the absence of collisions, the heterogeneous broadening during the different time moments and at the different energy transitions is fully correlated, since the frequency shifts are proportional to each other at different transitions and at different time intervals due to particle motion:

$$\Delta_j=\Omega_{0ik}\frac{\vec{v}_j\vec{n}}{c}=\Omega_{0ik}\frac{v_j\cos\varphi_j}{c}, \tag{2}$$

where Δ_j is the frequency shift due to the motion of the j-th particle with a velocity of $\vec{v}_j$ (in the absence of collisions), Ω_{0ik} is the central frequency of the i-k transition, and φ_j is the angle between the molecular velocity and the direction of observation

Let us consider the situation where the observations of a gas emission are made at different time moments separated by an interval τ at which elastic collisions of gas molecules can occur, leading to a change in the direction of the particles' velocity. These collisions would result in random changes in the projection of the velocity on the direction of the observation and, accordingly, random changes in frequency shifts (2) at each individual collision:

$$\Delta_j\left(\varphi_j,\delta\varphi_\eta(\Delta)\right)=\Omega_{0ik}\frac{v_j\cos\left(\varphi_j+\delta\varphi_\eta(\Delta)\right)}{c}, \tag{3}$$

where $\delta\varphi_\eta(\Delta)$ is the change in the magnitude of the angle between the direction of the velocity of the molecule and the direction of observation due to a random change in the direction of the velocity of the molecule as a result of an individual η-th collision. Thus, each isochrome of the Doppler-broadened line breaks down over the time as a result of the spectral diffusion. This leads to non-correlation of the heterogeneous broadening in the gas at different moments of time and partial loss of the phase memory of the system.

Let us find the average number of collisions at the time interval τ between the moments of observation of the system and their effect on the radiation frequency of the molecule. Over a second the molecule will, on average, pass a path equal to the average speed. Collisions with the change in velocity "distort" its path. We denote the effective diameter of the molecule by d and represent the molecule as a sphere. Then, the number of collisions of N_0 molecule with other molecules per second will be equal to the number of the molecules whose centers are in a cylinder of length equal to the mean velocity and the diameter of $2d$ [7]:

$$N_0 = \pi d^2 \bar{v} \rho,$$

where ρ is the number of molecules per unit volume. But we need to make a correction to the fact that this molecule collides with not immobile molecules, but with moving molecules [1]. This circumstance will be taken into account if the average relative velocity is used instead of the mean absolute velocity. Consequently, $\upsilon_r \approx \sqrt{2}\bar{v}$. Therefore:

$$N_0 = \sqrt{2}\pi d^2 \bar{v} \rho.$$

Over the time interval τ between the moments of the observation time of the system, an average of $N(\tau) = N_0\tau = \sqrt{2}\pi d^2 \bar{v}\rho\tau$ collisions will occur. This will lead to a change in the angle between the velocity of the molecule and the direction of observation in (3):

$$\Delta_j(\varphi_j, \delta\varphi_\eta(\Delta)) = \Omega_{0ik} \frac{v_j \cos\left(\varphi_j + \sum_{\eta=1}^{N(\tau)} \delta\varphi_\eta(\Delta)\right)}{c}, \tag{4}$$

where the number of collisions is

$$N(\tau) = \frac{\bar{v}}{\bar{l}}\tau,\ \bar{v} = \sqrt{\frac{8kT}{\pi m}}.$$

3. The impact of collisions on the temporal shape of the stimulated echo-hologram in a gas

The described above spectral diffusion within the heterogeneously broadened line influences the temporal form of the SEH response [6]. Let us consider the case, where the information is included into the temporal form of the second laser pulse when writing SEH. In this case, the pulse sequence for excitation of a stimulated echo hologram is shown in figure 1.

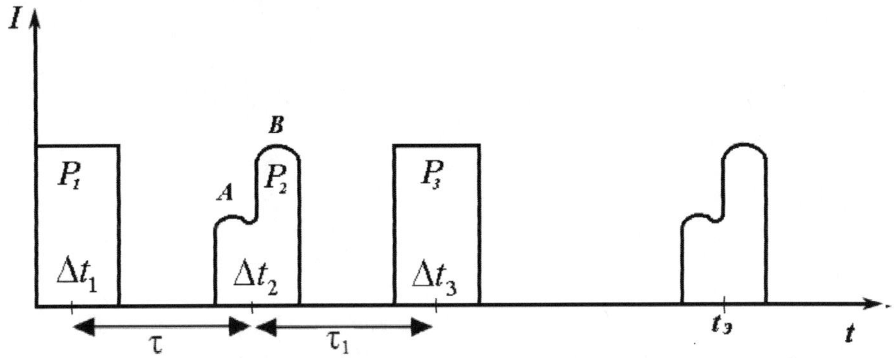

Figure 1. The sequence of pulses at SEH excitation. Δt_i is the duration of the i-th resonant laser pulse, t_e is the time of appearance of the SEH response, $\tau = t_2 - t_1$ and $\tau_1 = t_3 - t_1$ are the time intervals between the exciting pulses, and P_1, P_2, P_3 are the resonant laser pulses.

The electric field strength of the SEH response with allowance for (1-4) in case of linear polarization of the exciting laser pulses will be determined by an expression similar to that obtained in [8-9]:

$$E \sim \int_{-\infty}^{\infty} \frac{\sin(\theta_1(\Delta))\sin(\theta_2(\Delta,\delta\varphi_\tau(\Delta)))\sin(\theta_3(\Delta,\delta\varphi_{\tau_1}(\Delta)))}{\theta_1'(\Delta)\theta_2'(\Delta,\delta\varphi_\tau(\Delta))\theta_3'(\Delta,\delta\varphi_{\tau_1}(\Delta))} \times$$
$$\tilde{S}_1^*(\Delta)\tilde{S}_2(\Delta)\tilde{S}_3(\Delta)\exp\{i\Delta[t\cdot\cos(\delta\varphi_t(\Delta)) - \tau\cdot\cos(\delta\varphi_\tau(\Delta)) - \tau_1\cdot\cos(\delta\varphi_{\tau_1}(\Delta))]\}g(\Delta)d\Delta, \quad (5)$$

where

$g(\Delta) = \left(1/\Delta_D\sqrt{2\pi}\right)\exp\left\{-(\Delta/\Delta_D)^2\right\}$, $\theta_1(\Delta) = \theta_1'(\Delta)\delta t_1$, $\theta_2(\Delta,\delta\varphi_\tau(\Delta)) = \theta_2'(\Delta,\delta\varphi_\tau(\Delta))\delta t_2$, $\theta_3(\Delta,\delta\varphi_{\tau_1}(\Delta)) = \theta_3'(\Delta,\delta\varphi_{\tau_1}(\Delta))\delta t_3$ are areas of laser pulses, $\tilde{S}_1(\Delta)$, $\tilde{S}_2(\Delta)$, $\tilde{S}_3(\Delta)$ are Fourier spectra of pulses, $\delta\varphi_\tau(\Delta)$, $\delta\varphi_{\tau_1}(\Delta)$, $\delta\varphi_t(\Delta)$ is the average random increase in the angle between the velocity of the molecule and the direction of observation during corresponding time interval and the corresponding isochrome of the Doppler-broadened line, Δ_D is the width of the heterogeneously (Doppler) broadened line, τ is the time interval between the first and the second pulses, τ_1 is the time interval between the 1st and 3rd pulses.

Numerical calculation of expression (5) in case of information encoding into temporal form of laser pulse for the SEH response is shown in figure 2 (a, b) in cases where the time intervals have been changed between the 2nd and 3rd excitation laser pulses at different vapor pressures of ^{174}Yb.

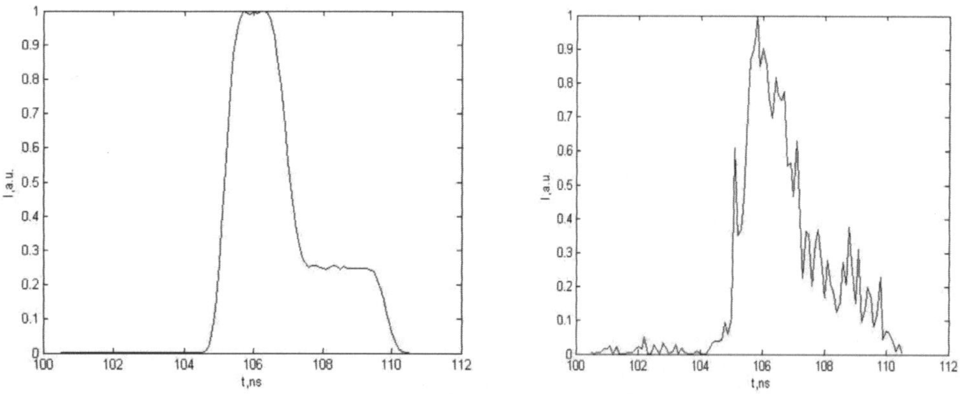

a) b)

Figure 2. The temporal form of the stimulated photon echo response in the gas: $\tau = 5$ ns, $\tau_1 = 100$ ns, $T_2 = 1000$ ns, $T = 800$ ° K, $d = 0.388\cdot10^{-7}$ cm, a) P = 1.5 Torr, b) P = 5.25 Torr.

4. Conclusions

Collisions with the change in velocity direction of gas particles lead to the destruction of the temporal form of the SEH response. Therefore, the reproducibility of the information in the SEH response due to collisions with the change in speed will deteriorate, which must be taken into account when creating an optical memory and information processing systems in gas media.

Acknowledgments
The work is performed according to the Russian Government Program of Competitive Growth of Kazan Federal University.

IOP Conf. Series: Journal of Physics: Conf. Series **859** (2017) 012002 doi:10.1088/1742-6596/859/1/012002

References

[1] Nefed'ev L, Samartsev V 1992 *J. Laser Phys.* **2** 617
[2] Rubtsova N, Goldort V, Ishchenko V, Kochubei S, Khvorostov E and Yevseyev I 2008 *JETP Lett.* **87** 103
[3] Evseev I, Ishchenko V, Kochubei S, Rubtsova N, Khvorostov E and Reshetov V 2006 *Bulletin of the Russ. Academy of Sci.: Phys.* **70** 527
[4] Rubtsova N, Ishchenko V, Khvorostov E, Kochubei S, Khvorostov E and Yevseyev I 2004 *J. Phys. Rev.* **70** 023403
[5] Popov I, Vashurin N, Gazizov K, Stepanov S, Sidorova V, Putilin S 2012 *Bulletin of the Russ. Academy of Sci.: Phys.* **76** 279
[6] Akhmedshina E, Nefediev L and Garnaeva G 2015 *J. of Appl.Spectr.* **82** 669
[7] Kikoin A and Kikoin I 1976 *Molecular physics* (Moscow: Science) p 480
[8] Evseev I, Rubtsova N and Samartsev V 2009 *Photon echo and phase memory in gases* (Kazan: Kazan State University Press) p 490
[9] Evseev I, Rubtsova N and Samartsev V 2009 *Coherent transient processes in optics* (Moscow: OOO Publishing company "Physical and mathematical literature") p 536

Evolution of a two-level quantum particle in the noise classical e.-m. field within and beyond the resonant approximation

A M Basharov

Moscow Institute of Physics and Technology, 141700 Dolgoprudny, Russia
National Research Centre «Kurchatov Institute», 123182 Moscow, Russia

E-mail: basharov@gmail.com; http://basharov.me

Abstract. The paper has shown that the interaction of a stochastic field with a two-level system beyond the resonant approximation is equivalent to the resonant interaction of a noise field in maximally squeezed state.

1. Introduction

In some cases interaction of delta-correlated random fields with quantum systems allows for a fully analytical analysis based on stochastic differential equations (SDE). We differentiate between resonant and non-resonant interaction models. With a two-level quantum system, resonance approximation is also called the rotating wave approximation and its corrections are considered - the so-called counter-rotating (anti-rotating) terms [1]. Using a random delta-correlated field allows considering the two cases analytically and demonstrating that the choice of a particular approach is determined by the nature of the random field - the ratio between the scale of the random change of a physical quantity represented as a delta-correlated field and the reverse transition frequency between a pair of the observed energy levels of a quantum particle. The below analysis for the classic delta-correlated field shows that from the standpoint of the kinetic equation for the density matrix of the quantum system, both cases can be presented in a unified manner in the framework of the resonance approximation, which uses different views on the squeezing of a random field.

2. Interaction of a random field with a two-level quantum particle by the Wiener process

Let a two-level atom with Hamiltonian H_a and energy levels $|E_1>$ and $|E_2>$ interact in the electric dipole approximation with the electromagnetic field intensity $\vec{E}$, so that the model of such interaction is based on the following Schrödinger equation for the wave function $|\Psi>$ of the whole system in the interaction picture

$$i\hbar\frac{d}{dt}|\Psi(t)>=H(t)|\Psi(t)>,\ H(t)=-\vec{E}(t)\vec{d}(t),\ |\Psi(t)>=U(t)|\Psi(0)>,$$

$$i\hbar\frac{d}{dt}U(t)=H(t)U(t),\ U(0)=1. \tag{1}$$

Here $\vec{d}(t)=\sum_{jk}\vec{d}_{jk}e^{i\Omega_{jk}t}\,|E_j><E_k|$ is the vector dipole moment operator of the atom. An atomic subsystem is represented by projection operators $|E_i><E_j|$ onto the atomic states $|E_i>$, described as energies E_i and orthogonal expansion of unity:

$$\sum_i |E_i><E_i|=1,\ <E_i|E_j>=\delta_{ij},\ H_a=\sum_i E_i|E_i><E_i|.$$

The transition frequency between the levels of a quantum particle is designated as $\Omega_{21}=(E_2-E_1)/\hbar>=-\Omega_{12}$.

The solution of equation (1) can be written in the form of series

$$U(t)=I+(-\frac{i}{\hbar})\int_0^t H(t')dt'+(-\frac{i}{\hbar})^2\int_0^t\int_0^{t'} H(t')H(t'')dt'dt''+...=\vec{T}\exp\left(-\frac{i}{\hbar}\int_0^t H(t')dt'\right). \quad (2)$$

Neglecting the polarization effects the interaction operator can be rewritten as

$$H(t)=-(d_{12}e^{-i\Omega_{21}t}R_-+d_{21}e^{i\Omega_{21}t}R_+)E(t),\ R_+=|E_2><E_1|,\ R_-=|E_1><E_2|=R_+^+.$$

A natural requirement imposed on $E(t)$ in the case of its representation in the form of a random process is stationarity of this value, i.e. the dependence of the correlation function $<E(t)E(t')>=g(t-t')$ only on the time difference $t-t'$.

When $E(t)$ is expanded in a Fourier integral, the stationarity of interaction results in the noncorrelation (independence) of spectral components

$$<b(\omega)b^*(\omega')>=\delta(\omega-\omega')S(\omega),$$

where $S(\omega)$ is the power spectrum of the random effects, with the Fourier components being introduced:

$$b(\omega)=\frac{1}{2\pi}\int e^{i\omega t}\xi(t)d\omega,\ b(\omega)=b^*(-\omega)\ \text{for}\ E(t)=E^*(t)=\int e^{-i\omega t}b(\omega)d\omega.$$

If we take the natural approach $S(\omega)=\text{const}$ for a specific spectral range and extend it to the whole region of the spectrum (Nyquist approximation), it is easy to see that

$$<E(t)E(t')>=D\delta(t-t').$$

Then the value $\int_0^t E(t')dt=W(t)$ is continuous and is nowhere differentiable. It can be shown in many ways that the density of the probability distribution function $\rho(w,t)$ of value $W(t)$ satisfies the kinetic equation (Fokker-Planck equation [2]) $\frac{\partial\rho(w,t)}{\partial t}=\frac{D}{2}\frac{\partial^2\rho(w,t)}{\partial w^2}$ with the solution $\rho(w,t|w_0,t_0)=\frac{1}{\sqrt{2\pi D(t-t_0)}}\exp[-\frac{(w-w_0)^2}{2D(t-t_0)}]$, for an initial condition $\rho(w,t_0|w_0,t_0)=\delta(w-w_0)$. The value $W(t)$ is called the Wiener process. The non-standard Wiener process is mentioned in certain circumstances, relating $D=1$ to the standard case.

Equation (1) is mathematically uncertain in the Nyquist approximation. But you can define strictly its integral "solution" (2). This requires a reasonable definition of the integral over the Wiener process $\int_{t_0}^{t} G(t')dW(t')$. A conventional definition of the integral $\int_{t_0}^{t} G(t')dW(t')$ as the limit of the partial sums $S_n = \sum_{i=1}^{n} G(\tau_i)(W(t_i) - W(t_{i-1}))$ leads to its dependence on the specific choice of intermediate points τ_i of the partition intervals: $t_{i-1} \le \tau_i \le t_i$. This can easily be seen from the example of the integral $\int_{t_0}^{t} W(t')dW(t')$ by using the properties of the Wiener process $W(t)$ - the statistical independence of its increments $W(t_i) - W(t_{i-1})$ of one another. The choice of Ito $\tau_i = t_{i-1}$ is convenient and self-consistent in case of the so-called non-anticipatory functions $G(t)$, statistically independent of the future behavior of the Wiener process at the time t. The limit of the integral sums is thought of as the mean square limit. Then you can introduce the established *Ito differentials* :

$$dW(t)dW(t) = Ddt\,,\ [dW(t)]^{N+2} = 0,\ N > 0,\ dW(t)dt = dtdt = 0\,,\ < dW(t) >= 0. \quad (3)$$

Here the angular brackets denote averaging and differential relations should be understood as fulfillment of integral equations for any nonanticipating functions such as:

$$\int_{t_0}^{t} G(t')dW(t')dW(t') = \int_{t_0}^{t} G(t')dt'.$$

Algebra (3) of Ito differentials allows obtaining SDE easily for the evolution operator

$$U(t) = \bar{T}\exp\left(i\int_{0}^{t} (d_{12}e^{-i\Omega_{21}t}R_- + d_{21}e^{i\Omega_{21}t}R_+)dW(t') \right).$$

Then

$$dU(t) = \{\exp(i(d_{12}e^{-i\Omega_{21}t}R_- + d_{21}e^{i\Omega_{21}t}R_+)dW(t')) - 1\}U(t).$$

Simple calculations based on Ito algebra (3) lead to the following SDE:

$$dU(t) = i(d_{12}e^{-i\Omega_{21}t}R_- + d_{21}e^{i\Omega_{21}t}R_+)dW(t)U(t) - \tfrac{1}{2}D(d_{12}e^{-i\Omega_{21}t}R_- + d_{21}e^{i\Omega_{21}t}R_+)^2U(t)dt\,, \quad (4)$$

In this case (as can be easily seen by direct calculation) the rule for Ito differentiating is also fulfilled

$$d(U^+(t)U(t)) = (dU^+(t))U(t) + U^+(t)dU(t)) + (dU^+(t))dU(t)\,,$$

and equations for the density matrix $\rho^a(t)$ of the atom are conventionally derived from the chain of equations:

$$d\rho^a(t) \equiv \rho^a(t+dt) - \rho^a(t)\,,$$
$$\rho^a(t+dt) = <|\Psi(t+dt)><\Psi(t+dt)| = U(t+dt)|\Psi(0)><\Psi(0)|U^+(t+dt)>,$$
$$d\rho^a(t) = <dU(t)|\Psi(0)><\Psi(0)|U^+(t) + U(t)|\Psi(0)><\Psi(0)|dU^+(t)$$
$$+ dU(t)|\Psi(0)><\Psi(0)|dU^+(t)>.$$

Upon averaging it is convenient to rewrite these equations in the following form

$$\frac{d\rho^a(t)}{dt} = n[2Y\rho^a(t)Y^+ - Y^+Y\rho^a(t) - \rho^a(t)Y^+Y] + m[2Y^+\rho^a(t)Y^+ - Y^+Y^+\rho^a(t) - \rho^a(t)Y^+Y^+]$$
$$+ n[2Y^+\rho^a(t)Y - YY^+\rho^a(t) - \rho^a(t)YY^+] + m^*[2Y\rho^a(t)Y - YY\rho^a(t) - \rho^a(t)YY]. \quad (5)$$

We introduce the notation

$$Y = d_{12}R_- ,\ Y^+ = d_{21}R_+ ,\ n = D ,\ m = De^{i2\Omega_{21}t} .$$

There exists the ratio $|m| = n$ between the introduced parameters n and m. However, below we can see that from the standpoint of the kinetic equation for the two-level atom, the description of the influence of the random field having neither a carrier frequency nor resonance approximation satisfies completely the kinetic equation derived in the resonance approximation under the influence of the resonance noise wave having a carrier frequency. In other words,

result (5) also describes the interaction with a two-level system in the resonance approximation of a noise wave in the maximally squeezed state with a carrier frequency, which coincides with the transition frequency between these levels.

3. Interaction of a random field with a two-level quantum particle in the rotating wave approximation

Now we are going to consider the interaction of a noise wave with the carrier frequency, which coincides with the transition frequency of a two-level system in the rotating wave approximation. The correct description of the noise wave with the carrier frequency suggests that we introduce a complex-valued Wiener process.

It is convenient to view the two non-standard independent Wiener processes $W_1(t)$ and $W_2(t)$ with algebra

$$dW_1(t)dW_1(t) = D_1dt ,\ dW_2(t)dW_2(t) = D_2dt ,\ dW_1(t)dt = dW_2(t)dt = dtdt = 0 .$$

They allow defining the complex-valued Wiener process as $\tilde{W}(t) = W_1(t) + iW_2(t)$, for which we have the following Ito algebra

$$d\tilde{W}(t)d\tilde{W}(t) = (D_1 - D_2)dt ,\ d\tilde{W}(t)d\tilde{W}^*(t) = (D_1 + D_2)dt ,\ d\tilde{W}(t)dt = d\tilde{W}^*(t)dt = dtdt = 0 .$$

Finally, the complex-valued Wiener process in a general form is defined as $\tilde{W}(t) = \tilde{W}_1(t) + e^{i\varphi}\tilde{W}_2(t)$, so

$$d\tilde{W}(t)d\tilde{W}^*(t) = (D_1 + D_2)dt \equiv ndt ,\ d\tilde{W}(t)dt = d\tilde{W}^*(t)dt = dtdt = 0 ,$$
$$d\tilde{W}(t)d\tilde{W}(t) = (D_1 + D_2e^{i2\varphi})dt \equiv mdt ,\ d\tilde{W}^*(t)d\tilde{W}^*(t) = (D_1 + D_2e^{-i2\varphi})dt \equiv m^*dt . \quad (6)$$

It is evident that

$$|m| \le n , \quad (7)$$

because

$$|D_1 + D_2e^{i2\varphi}| = \sqrt{(D_1 + D_2e^{i2\varphi})(D_1 + D_2e^{-i2\varphi})} = \sqrt{(D_1^2 + D_2^2 + 2D_1D_2\cos(2\varphi)} \le D_1 + D_2 .$$

Then you can introduce random processes with independent increments, satisfying conditions (6) with additional requirements (7) and

$$< d\tilde{W}(t) > = < d\tilde{W}^*(t) > = 0 .$$

IOP Conf. Series: Journal of Physics: Conf. Series **859** (2017) 012003 doi:10.1088/1742-6596/859/1/012003

Condition (7) is necessary in order that the components of the complex-valued process of this type $W_1(t)$ and $W_1(t)$ have a Gaussian distribution.

The introduction of complex-valued Wiener –type processes with algebra (6) - (7) is convenient to describe random electromagnetic fields with a carrier frequency. The complex-valued Wiener process in a general form allows you to write an interaction operator of the classical electromagnetic field with a two-level system in the rotating wave approximation in the form

$$H(t)dt = -d_{12}R_{-}dW^{*}(t') - d_{21}R_{+}dW(t').$$

We emphasize that this expression is meaningful only if the evolution operator of a quantum system is represented in the form of a T-exponent:

$$U(t) = \overleftarrow{T}\exp\left(i\int_0^t (d_{12}R_{-}dW^{*}(t') + d_{21}R_{+}dW(t'))\right).$$

Expanding the exponent in the expression $dU(t)$ we obtain

$$dU(t) = i(d_{12}R_{-}dW^{*}(t') + d_{21}R_{+}dW(t'))U(t)$$
$$-\frac{1}{2}(d_{12}^2 R_{-}R_{-}m^{*} + |d_{12}|^2 R_{-}R_{+}n + |d_{12}|^2 R_{+}R_{-}n + d_{21}^2 R_{+}R_{+}m)dtU(t).$$

As a result of the above described calculations we obtain the kinetic equation for the density matrix of a two-level atom in form (5), and the parameters n and m are not related to ratio $|m| = n$ as in the case of absence of any resonance approximation, but are governed by general equation (7) and defined by equations (6).

The case $m \neq 0$ corresponds to the so-called squeezing of an electromagnetic field whose creation requires some special conditions of modulation. The limiting case $|m| = n$ corresponds to the maximum squeezing, and with a two-level system suggests another interpretation outlined in the previous section.

Equality of the squeezing parameter to zero $m = 0$ describes the usual classical noise electromagnetic wave characterized by a carrier frequency. It is a case which corresponds to the modeling of resonant effects of a random field with a quantum system.

Thus, the resonance approximation or its rejection for the purpose of considering the alleged "general" case is practically determined by the parameters of the external influence. If the conditions of representation of external influence by a Wiener process are satisfied, then we obtain the solution of the equation for the evolution operator in the form of T-exponent, and derivation of the kinetic equation for the density matrix of a quantum system is an example of analytical calculations made with such expressions. If the resonance conditions are satisfied, it is necessary to develop a consistent algebraic perturbation theory [2], in the first order of which there arises the rotating wave approximation discussed above. If you restrict yourself to solving the problem at the level of the kinetic equation for the density matrix of a quantum system, then within the limits of the resonance approximation all the features of the noise field, including its squeezing, are considered, whereas beyond the resonance algebraic perturbation theory only the limiting case of the noise field state is described, but this restriction of the model of a multilevel quantum system by its two levels for a noise field without the carrier frequency is physically meaningless. It is also difficult to justify SDE when the characteristic time scale of random change in the field is of the order of the reverse frequency of quantum transition. For SDE to be justified, it is necessary that the scale of change in a random field were much less than the reverse frequency of quantum transition, which is unreal in the optical frequency range from the standpoint of physics of electromagnetic fields. However, in the case of a two-level system formed by sublevels of a quantum level, i.e. when the transition frequency of a two-level system is small, it is possible to consider the interaction process beyond the resonance approximation, provided that the other levels of a quantum system are either absent or located far

IOP Conf. Series: Journal of Physics: Conf. Series **859** (2017) 012003 doi:10.1088/1742-6596/859/1/012003

enough to apply the Nyquist approximation to the spectrum of a noise wave. Then formula (5), where $|m|=n$, leads to the kinetic equation for a quantum system beyond the resonance approximation.

Unfortunately, methods of the SDE are unpopular among scientists in spite of being very effective in analytical solution of many physical problems.

Acknowledgments
This work is supported in part by the Russian Foundation for Basic Research (grant No. 16-02-00453a).

4. References

[1] Mandel L and Wolf E 1995 *Optical Coherence and Quantum Optics* (New York: Cambridge University Press)
[2] Gardiner C W 2004 *Handbook of stochastic methods* (Berlin, Heidelberg: Springer-Verlag)
[3] Maimistov A I and Basharov A M 1999 *Nonlinear optical waves* (Dordrecht: Kluwer Academic)

IOP Conf. Series: Journal of Physics: Conf. Series **859** (2017) 012004 doi:10.1088/1742-6596/859/1/012004

Low-frequency emission in resonant processes

A M Basharov

Moscow Institute of Physics and Technology, 141700 Dolgoprudny, Russia
National Research Centre «Kurchatov Institute», 123182 Moscow, Russia

E-mail: basharov@gmail.com; http://basharov.me

Abstract. The paper shows analytically that any resonant interaction of quantum particles in media having no symmetry centre with a high-frequency exciting field produces low-frequency radiation. Media particles may not necessarily have a constant dipole moment. Conditions are identified when carrier frequency of low-frequency radiation coincides with nutational oscillations of resonant level populations and is readily controlled both by high-frequency resonant radiation intensity and by resonance detuning.

1. Introduction

In recent years generation of low-frequency radiation whose carrier frequencies are in the terahertz range has been actively explored with the purpose of practical application. Among the nonlinear-optical resonant phenomena, research works [1-5] studied resonant processes of interaction of optical (and other "high-frequency") waves with particles having permanent dipole moment. Research is focused not only on the generation of terahertz radiation, but also on the effect of permanent dipole moment on the excitation pulse propagation [6-9].

The paper has demonstrated analytically that for generation of low-frequency radiation in resonant-optical processes, including terahertz radiation, it is not necessary to study environments whose particles have permanent dipole moment. Generation of low-frequency radiation is also possible in ordinary environments in conditions of simple one-photon resonant interaction between exciting coherent pulses and an environment. It is necessary that resonant particles have non-resonant energy levels of indefinite parity, which may be formed, e.g., by levels with different parity but similar energy (which is characteristic of the media with quadratic nonlinear response). Meanwhile, in some certain conditions carrier frequency of low-frequency radiation precisely equals frequency of nutational oscillations of resonant level populations of quantum particles. Since frequency of nutational oscillations is proportional to the square root of resonance pumping intensity, it provides for a simple method of controlling carrier frequency of low-frequency radiation up to terahertz radiation generation.

As far as the intensity of generated low-frequency radiation is concerned, it proves to be as low as intensity of any two-quantum process in an electromagnetic wave.

The obtained results followed from considerations of resonant interactions up to the second-order perturbation theory. Second order of smallness is certain to be disadvantageous compared to the case of low-frequency generation for environments having particles with permanent dipole moment. At the same time, theoretical consideration of generation processes in environments containing resonant and populated levels with no permanent dipole moment is more correct since there arise no questions concerning phase transition in condensed environments whose particles have permanent dipole moments and which are not taken into account by low-frequency generation investigations. Both the intensity smallness of the explored radiation and the coincidence of its propagation direction with an exciting pulse direction is likely to hinder experimental detecting this radiation against powerful pumping.

Studying resonant interaction of coherent radiation with matter in theory and experiment, the so-called optical nutation [10,11] is referred to as one of the fundamental effects. Nutational oscillations are meant as oscillations of radiation intensity, having passed through optically thin resonant environment near its front edge if the exciting pulse envelope has a squared shape. At the same time, radiation intensity is considered to be averaged over a period of fast oscillations of an electromagnetic field. Frequency of nutational oscillations, multiplied by Planck constant, represents interaction energy of electromagnetic radiation with a resonant particle. Therefore, a coherent field with a squared shape envelope passing through the optically thin resonance environment has two frequency characteristics: radiation carrier frequency and quantum reradiation frequency within the field of this radiation (nutation), with radiation carrier frequency being far higher than that of nutational oscillations.

The paper has shown how radiation of carrier frequency, being equal to frequency of nutational oscillations, is formed under certain conditions in environments with no permanent dipole moment. Generally, it is not surprising because the paper defines that in a resonant wave field the particle dipole moment contains intermodulation components. A consistent theory of resonant interaction of coherent radiation with the particle environment with account of non-resonant particle states has been introduced; common expressions for effective operators of dipole moment of environmental particles have been derived. Application of the developed theory is a new possibility of controlling carrier frequency of the generated radiation by means of intensity change of an exciting wave. In particular, the considered process can be useful for generating quasi-monochromatic terahertz (or subterahertz) radiation.

2. Algebraic perturbation theory

Let propagate a coherent wave towards the axis Z whose intensity E of electric field is represented as a slowly changing amplitude $\mathcal{E}$ and fast changing phase defined by a wave vector $\vec{k} \parallel Z$ and carrier frequency ω:

$$E = \mathcal{E}\exp[i(kz - \omega t)] + c.c. \equiv \mathcal{E}\exp[-i\Phi] + c.c.$$

Polarization effects will be neglected assuming that electric vector is parallel to axis X. Letters $c.c.$ designate terms complexly conjugated with the previous term. $\Phi = \omega t - kz$ is the phase of a coherent wave.

Semispace $Z \geq 0$ is filled with environment consisting of resonant particles. Particle resonance levels consist of ground state $|g>$ and excited state $|e>$. Frequency ω_{eg} of transition $|e> \to |g>$ is close to frequency ω, $|\omega_{eg} - \omega| << \omega$. Generally, in analogous problems transition $|e> \to |g>$ is referred to as optically allowed transition supposing, firstly, that dipole transition moment $d_{eg} \neq 0$, and, secondly, that states $|e>$ и $|g>$ are characterized by different parity. These suppositions may also be accepted in our consideration, however, requirement for different parity may be omitted – it is enough that only $d_{eg} \neq 0$ exists, but permanent dipole moment is absent $d_{ee} = d_{gg} = 0$.

Neglecting interaction between particles, dipole moment of environment unit volume P can be expressed as

$$P = nSp\rho d,$$

where n - is the density of resonance particles per unit volume, $d = \sum_{kj} d_{kj} | E_k >< E_j |$ is the operator of dipole moment of resonance particle, $| E_k >$ is the quantum state of resonance particle, which may be degenerated, and a ρ is its density matrix. Polarization of environment P enters Maxwell equation

$$(\frac{\partial^2}{\partial z^2} - \frac{1}{c^2}\frac{\partial^2}{\partial t^2})E = \frac{4\pi}{c^2}\frac{\partial^2}{\partial t^2}P .$$

In ignoring relaxation and spontaneous particle radiation as well as with account of electrodipole approximation in interaction of particle with a coherent field, equation for density matrix takes a standard form

$$i\hbar\frac{\partial\rho}{\partial t} = [H_0 - Ed, \rho] .$$

Following the general method [10], let us transform density matrix:

$$\tilde{\rho} = U\rho U^+ ,\ U = e^{-iS} ,\ S^+ = S .$$

Transformed density matrix will satisfy kinetic equation

$$i\hbar\frac{\partial\tilde{\rho}}{\partial t} = [\tilde{H}, \tilde{\rho}]$$

with the transformed Hamiltonian

$$\tilde{H} = U(H_0 - Ed)U^+ - i\hbar U \tfrac{\partial}{\partial t} U^+ .$$

Such representation is convenient for developing perturbation theory. Let us expand $\tilde{H}$ and S into a series according to the interaction with a coherent field, marking with a superscript a size of order of terms with regard to coupling constant of the coherent field:

$$S = S^{(1)} + S^{(2)} + \ldots,\ \tilde{H} = \tilde{H}^{(0)} + \tilde{H}^{(1)} + \tilde{H}^{(2)} + \ldots$$

Taking the Baker–Hausdorff formula [11] into account, we obtain

$$\tilde{H}^{(0)} = H_0 ,\ \tilde{H}^{(1)} = -Ed - i[S^{(1)}, \tilde{H}^{(0)}] + \hbar\partial S^{(1)} / \partial t,$$

$$\tilde{H}^{(2)} = \frac{i}{2}[S^{(1)}, Ed] - \frac{i}{2}[S^{(1)}, \tilde{H}^{(1)}] - i[S^{(2)}, \tilde{H}^{(0)}] + \hbar\frac{\partial S^{(2)}}{\partial t} .$$

The main idea to find terms $S^{(i)}$ and the effective Hamiltonian $H^{Eff} = \tilde{H}^{(0)} + \tilde{H}^{(1)} + \tilde{H}^{(2)}$ serves the absence of fast-changing in time terms in $\tilde{H}^{(i)}$ in interaction representation. This leads to "correctly" changing in time terms in the following expressions:

$$\tilde{H}^{(1)} = -\mathscr{E}\exp(-i\Phi)d_{ge} | E_g >< E_e | + H.c.,\ S^{(1)} = \sum_{jk}{}' \frac{d_{jk}}{i\hbar}\left(\frac{\mathscr{E}\exp(-i\Phi)}{\omega_{jk} - \omega} + \frac{\mathscr{E}^*\exp(i\Phi)}{\omega_{jk} + \omega}\right) | E_j >< E_k | ,$$

$$\tilde{H}^{(2)} = -|\mathscr{E}|^2 \sum_k \Pi_{kk}(\omega) | E_k >< E_k | ,$$

where main parameters of the theory of optical resonant processes [11] are involved

$$\Pi_{kj}(\omega)=\sum_j{}'\frac{d_{kn}d_{nj}}{\hbar}\left(\frac{1}{\omega_{nk}+\omega}+\frac{1}{\omega_{nj}-\omega}\right)=\Pi_{jk}{}^*(-\omega).$$

Letters $H.c$ designate terms which are Hermitean-conjugated to the previous terms. Stroke of the summation symbol indicates that resonant terms $\omega_{jk}\mp\omega\approx 0$ are excluded in summation.

It is important to stress that $\Pi_{eg}(\omega)\neq 0$ for $|e>$ and $|g>$ states, even in the case of different parity of resonant states $|e>$ и $|g>$, since in summation in expression for $\Pi_{eg}(\omega)$ all states are accounted, including degenerate states and states with no certain parity. Moreover, the theory does account for both certain parity and absence of parity of resonant states.

Polarization of environment, expressed in terms of the transformed density matrix $P=nSp\tilde{\rho}\tilde{d}$, is defined by the effective operator of dipole moment

$$\tilde{d}=e^{-iS}de^{iS}=d-i[S,d]-\frac{1}{2}[S,[S,d]]+...=\sum_{kj}\tilde{d}_{kj}\,|E_k><E_j|,$$

$$\tilde{d}_{kj}\approx d_{kj}+\mathcal{E}\Pi_{kj}(\omega)e^{-i\Phi}+\mathcal{E}^*\Pi_{kj}(-\omega)e^{i\Phi}. \quad (1)$$

Characteristic feature of polarization of resonant environment with $\Pi_{eg}(\omega)\neq 0$, but in absence of permanent dipole moment $d_{ee}=d_{gg}=0$, is the presence of several characteristic frequencies in the field of resonant pulse ($\tilde{\rho}_{eg}=R_{eg}e^{-i\Phi}$, R_{eg} is the slowly changing in time amplitude of coherence of a resonant particle):

$$\begin{aligned}P=n\tilde{\rho}_{eg}\tilde{d}_{ge}+H.c.=nR_{eg}d_{ge}e^{-i\Phi}+n\mathcal{E}R_{eg}\Pi_{ge}(\omega)e^{-2i\Phi}+n\mathcal{E}^*R_{eg}\Pi_{ge}(-\omega)+\\+\mathcal{E}R_{ee}\Pi_{ee}(\omega)e^{-i\Phi}+\mathcal{E}R_{gg}\Pi_{gg}(\omega)e^{-i\Phi}+H.c.\end{aligned} \quad (2)$$

Among them the term of interest, which takes the following form

$$P_\Lambda=n\mathcal{E}^*R_{eg}\Pi_{ge}(-\omega)+H.c.$$

is characterized only by frequencies of slowly changing coherence amplitude R_{eg}.

3. Generation of low-frequency radiation in proximity of front edge of resonant pulse

Let the coherent pulse entering the resonant environment be rectangular in shape ($a=a^*$):

$$\mathcal{E}(z=0,t)=\begin{cases}a,t\geq 0,\\0,t<0.\end{cases}$$

Then for optically thin environments of length L, when the feedback effect on the passing pulse can be neglected, amplitude of coherence of resonant particles with resonance detuning Δ takes the form

$$R_{eg}=\frac{ad_{eg}/\hbar}{\Omega}[\frac{\Delta}{\Omega}(1-\cos\Omega t)+i\sin\Omega t],$$

$\Omega=\sqrt{\Delta^2+\Lambda^2}$, $\Lambda=2|ad_{eg}/\hbar|$. Let us suppose that distribution of resonant particles throughout detunings from rigid resonance is given by some function $F(\Delta)$.

The considered amplitude R_{eg} of coherence of resonant atoms provides for oscillations of environment polarization whose central frequency coincides with frequency Λ.This is exactly the frequency of nutational oscillations of resonant level populations.

IOP Conf. Series: Journal of Physics: Conf. Series **859** (2017) 012004 doi:10.1088/1742-6596/859/1/012004

$$P_{\Lambda} = n\Pi_{ge}(-\omega)\frac{a^2 d_{eg}}{\hbar} < \frac{1}{\Omega}[\frac{\Delta}{\Omega}(1-\cos\Omega t) + i\sin\Omega t] > +H.c. \quad (3)$$

Angular brackets indicate averaging over spread of detuning with a distribution density $F(\Delta)$.

Thus, periodic dynamics of quantum populations over Rabi frequency gives rise (by means of a term in an effective dipole moment of quantum particle) to polarization oscillations over nutation frequency. In order to find intensity of electric field E_{Λ}, emitted by these polarization oscillations, it is necessary to solve the Maxwell equation.

Radiation over frequency of nutational oscillations of resonant levels is derived from approximation of unidirectional waves [12]. Then the Maxwell equation is reduced to [12]:

$$(\frac{\partial}{\partial z} + \frac{1}{c}\frac{\partial}{\partial z})E_{\Lambda} = -\frac{2\pi}{c}\frac{\partial}{\partial t}P_{\Lambda}.$$

For optically thin environments of length L we obtain that beyond the limits of slowly changing electric field amplitude, generated at low frequency, takes the form:

$$E_{\Lambda} = -\frac{1}{2}\pi n L\hbar c^{-1}\frac{\Pi_{ge}(-\omega)}{d_{ge}}\Lambda^2 < \frac{\Delta}{\Omega}\sin\Omega(t-\frac{z}{c}) + i\cos\Omega(t-\frac{z}{c}) > +c.c. \quad (4)$$

Broadly speaking, parameters of resonant environment entering coefficient $\Pi_{ge}(-\omega)/d_{ge}$, are not purely real values. So

$$\Pi_{ge}(-\omega)/d_{ge} = \mathrm{Re}(\Pi_{ge}(-\omega)/d_{ge}) + i\,\mathrm{Im}(\Pi_{ge}(-\omega)/d_{ge}).$$

It follows from formula (4) that central frequency of low-frequency oscillations coincides with frequency of nutational oscillations of resonant level populations.

At resonance detuning with no energy level spread (or at narrow frequency spread compared with Λ) and $\mathrm{Im}(\Pi_{ge}(-\omega)/d_{ge}) = 0$:

$$E_{\Lambda} = -\pi n L\hbar c^{-1}\frac{\Pi_{ge}(-\omega)}{d_{ge}}\Lambda^2\frac{\Delta}{\Omega}\sin\Omega(t-\frac{z}{c}).$$

Frequency of low-frequency oscillations coincides with frequency of nutational oscillations.

Frequency spread of resonant particles leads to attenuation of low-frequency radiation despite the fact that its exciting radiation continues affecting resonance environment. In addition, in the simplest case of exact resonance $\Delta = 0$ and the pure real meaning of value $\Pi_{ge}(-\omega)/d_{ge}$ the low-frequency radiation is absent $E_{\Lambda} = 0$. This fact, along with coincidence of direction of studied low-frequency radiation and direction of pumping resonant wave, hinders experimental observation of low-frequency radiation (4).

Practical application of studied low-frequency radiation is the possibility of tuning of radiation carrier frequency by means of change in pumping intensity. For frequency of low-frequency radiation under discussion to be in terahertz region $\Lambda \sim 10^{12}$ s^{-1} (what is required in many problems) exciting resonant fields of order $I \sim 10^8$ Wt/cm^2, are needed if one estimates that $d_{eg} \sim 10^{-18}$ CGSE and takes into account the following expression $\Lambda = (8\pi I c^{-1}\hbar^{-2})^{1/2} d_{eg}$.

4. Conclusion

The result under discussion has so far not received due attention in theoretical research works on nonlinear-optical events in resonant environments. Obviously, it is connected with the fact that processes of environment resonance interaction are treated by researchers in terms of two-level models constructed not on the basis of the strict perturbation theory but merely by suppression of non-resonant levels [13,14].

To interpret two-level models physicists generally use averaging methods while considering terms that change fast and slowly in time (such as Bogoliubov-Mitropolsky method [15,16]). But in case your attention is focused only on resonant effects in the averaging method it is hard to notice, due to its awkwardness, other parametric processes generated by resonant interaction.

The averaging method provides for correct effective equations. However, in case of optical problems it is difficult to present the whole picture of parametric interactions of waves in a resonant environment in a simple and vivid way, using the above indicated method. In this paper a more simple and effective approach is based on algebraic perturbation theory [17,18], which was developed in works [11,19] as applied to nonlinear-optical problems.

Consistently applied, the theory of resonant interaction in terms of algebraic perturbation theory [11] allows not only explain rigidly the approximation of two-level models but also in a simpler way, compared to averaging methods, to find expressions for theory parameters which possess characteristics of all non-resonant levels of particles in resonant environment along with characteristics of all resonant levels. In addition, the method allows embracing the variety of parametric interactions in resonant environment and identify the described effect of low-frequency radiation at the frequency of nutational oscillations of resonant level populations of environment.

In conclusion, it should be noted that the phase switching method applied to the exciting high-frequency field [20] can be used to provide for an effective resonant interaction of the medium with an exciting field of a rectangular shape.

Acknowledgments

This work is supported in part by the Russian Foundation for Basic Research (grant No. 16-02-00453a).

References

[1] Sazonov S V and Ustinov N V 2005 *Quantum Electronics* **35** 701
[2] Bugai A N and Sazonov S V 2010 *Phys. Lett.* A **374** 1093
[3] Bugai A N and Sazonov S V 2014 *JETP Lett.* **98** 638
[4] Kmetic M A and Meath W J 1985 *Phys. Lett.* A **108** 340
[5] Basharov A M 2016 *JETP Lett.* **103** 15
[6] Maimistov A I and Caputo J-G 2003 *Optics and Spectroscopy* **94** 245
[7] Agrotis M, Ercolani N M, Glasgow S A and Moloney J V 2000 *Physica* D **138** 134
[8] Glasgow S A, Agrotis M A and Ercolani N M 2005 *Physica* D **212** 82
[9] Zabolotskiĭ A A 2008 *JETP* **106** 846
[10] Allen L and Eberly J H 1975 *Optical resonance and two-level atoms* (New York: Wiley–Interscience)
[11] Maimistov A I and Basharov A M 1999 *Nonlinear optical waves* (Dordrecht: Kluwer Academic)
[12] Belenov E M and Nazarkin A V 1990 *JETP Lett.* **51** 288
[13] Mandel L and Wolf E 1995 *Optical coherence and quantum optics* (Cambridge University Press)
[14] Weiner J and Ho P-T 2003 *Light-matter interaction* (New York: Wiley–Interscience)
[15] Bogolubov N N and Mitropolsky Yu A 1963 *Asymptotic methods in nonlinear oscillation theory* (Moscow: Nauka).
[16] Grebenikov E, Mitropolsky Yu A and Ryabov Y A 2004 *Asymptotic methods in resonance analytical dynamics* (CRC Press)
[17] Bogaevski V N and Povzner A 1991 *Algebraic methods in nonlinear perturbation theory* (New-York: Springer-Verlag)

IOP Conf. Series: Journal of Physics: Conf. Series **859** (2017) 012004 doi:10.1088/1742-6596/859/1/012004

[18] Adams B G 1994 *Algebraic Approach to simple quantum systems. With applications to perturbation theory* (Berlin, Heidelberg: Springer-Verlag)
[19] Basharov A M, Maimistov A I and Manykin E A 1983 *Sov.Phys.JETP* **57** 282
[20] Alekseev A I and Basharov A M 1982 *J. Phys.* B **15** 4269

Investigation of the domain structure transformation under mechanical deformations in permalloy microparticles

D A Bizyaev[1], A A Bukharaev[1,2], N I Nurgazizov[1,2] and T F Khanipov[1]

[1]Kazan E. K. Zavoisky Physical -Technical Institute, 420034, 10/7 Sibirskiy Trakt, Kazan, Russian Federation
[2]Kazan Federal University, 420008, 18 Kremlevsakya St, Kazan, Russian Federation

E-mail: a_bukharaev@mail.ru

Abstract. Using magnetic force microscopy (MFM) and computer simulation it was shown that the mechanical compression of the permalloy microparticles leads to the increase in the effective anisotropy field and the noticeable decrease in the external magnetic field value necessary for the formation of the uniform magnetization in the compressed particle. The analysis of MFM images of microparticles covering the whole substrate surface made it possible to conclude about the uniform or nonuniform distribution of stresses induced in the particles in the different area of the substrate.

1. Introduction

At present the processes occurring in magnetic media under various external impacts are studied intensively [1]. This is associated with the fact that mechanical stress can decrease significantly the energy of one bit writing in magnetic data storage media. The study of processes of the magnetic anisotropy variation in the magnetostriction layer under the mechanical deformation and the possibility of using these processes to fabricate logic gates for micro- and nanoelectronics with the extremely low energy consumption led to the appearance of a new field of science – straintronics [2]. In addition, the Villary effect (the change of magnetic properties of the material under mechanical stress) can be used for the visualization of mechanical stress, since in the first approximation the induced magnetic anisotropy is proportional to the stress tensor [3].

Recently using magnetic force microscopy (MFM) we found out that the domain structure of square permalloy (Py) microparticles changes noticeably due to the mechanical compression [4, 5]. Here this effect is used for estimating the uniformity rate of the induced stress distribution in the particles covering the preliminarily bent substrate after straightening. In addition, the effect of mechanical compression on the formation of the uniform magnetization in Py microparticles under the external magnetic field is studied using MFM.

2. Results and discussion

Square Py microparticles with dimensions of $25\times25\times0.03$ μm^3 and $7\times7\times0.03$ μm^3 were formed for the studies on hard substrates. Particles were fabricated using the electron beam Py target evaporation through the grid with the appropriate hole size. The Py (Ni 75%, Fe 25%) layer with the thickness of 30 nm was deposited in ultrahigh vacuum. A Multiprobe P (Omicron) device was used for the sample preparation.

IOP Conf. Series: Journal of Physics: Conf. Series **859** (2017) 012005 doi:10.1088/1742-6596/859/1/012005

Particles with dimensions of 7×7×0.03 μm^3 were fabricated on the Si substrate with dimensions of 10×5×0.35 mm^3. Samples with unstressed particles and samples with uniformly stressed particles along one of the substrate axis were fabricated. Uniformly stressed samples were prepared using a special method. At first the Si substrate was bent elastically in a special holder (figure 1f) prior to the Py target evaporation. The constant curvature radius of the holder makes it possible to strain the substrate uniformly during the Py deposition. After the deposition, sample was extracted from the holder, the substrate was unbent and particles were compressed uniformly on the substrate surface along one of the axis. MFM measurements of the fabricated Py particle array were carried out by Solver P47 Pro (NT-MDT) scanning probe microscope (SPM) using N18/Co-Cr (MikroScience) magnetic cantilevers. The magnet integrated to SPM gave the possibility to vary the magnetic field in the plane of the sample up to 200 Oe.

According to MFM measurements, unstressed particles had the four-domain magnetic structure with single-size domains (figure 1a). Stressed particles also had the four-domain structure. However, domains with the magnetization perpendicular to the direction of the compression were larger than those with the magnetization parallel to the direction of the compression (figure 1d). The characteristic bridge was formed between two large domains. According to our studies, the length of the bridge depends on the mechanical stress value in the particle and it can be used as a parameter to characterize the compression rate [4, 5]. MFM measurements of the stressed sample showed that almost all microparticles had the bridge with the same size (figure 1d). It indicates the uniformity of the particle compression over the sample.

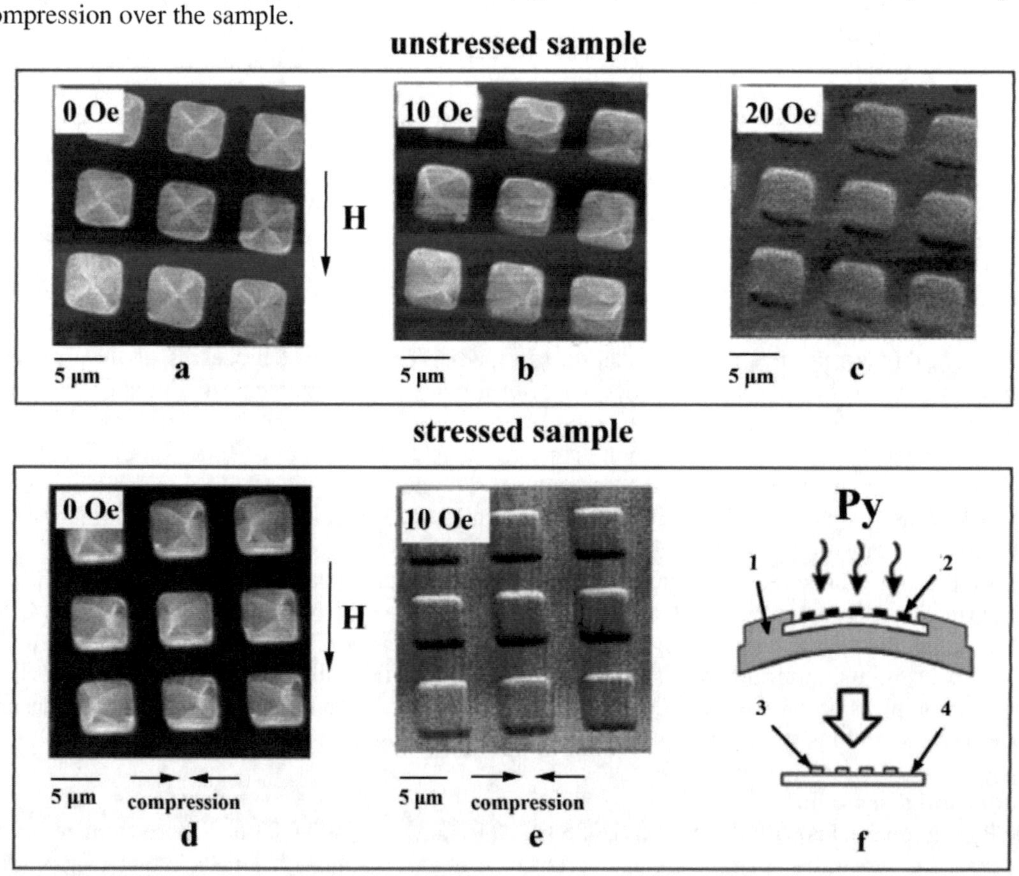

Figure 1. MFM scans of unstressed Py particles with dimensions of 7×7×0.03 μm^3 in the external magnetic field: a) 0, b) 10, c) 20 Oe. MFM scans of stressed Py particles with dimensions of 7×7×0.03 μm^3 in the external magnetic field: d) 0, e) 10 Oe. Scheme of the preparation of a stressed sample – f (1 – holders; 2 – metal grid; 3 – Py microparticles, 4 – Si substrate).

The formation of the uniform magnetization in uncompressed and compressed microparticles with dimensions of 7×7×0.03 μm^3 in the external magnetic field applied in the plane of the sample surface was studied using an SPM with the integrated magnet (figure 1b, c, e). When the uniform magnetization state is formed in microparticles, the domain walls vanish and their MFM images have the characteristic shape owing to the formation of two magnetic poles on opposite sides of microparticles (figure 1c, e). The experiments showed that the uniform magnetization state is formed in the field of 10 Oe for compressed particles, while for uncompressed particles this field is twice as large (figure 1c,e). This result is in agreement with result [6], where it was shown that the combination of the mechanically induced stress and magnetic field makes it possible to form the uniform magnetization in the given direction in ferromagnetic microparticle. This effect can be used for straintronics magnetic data storage [2].

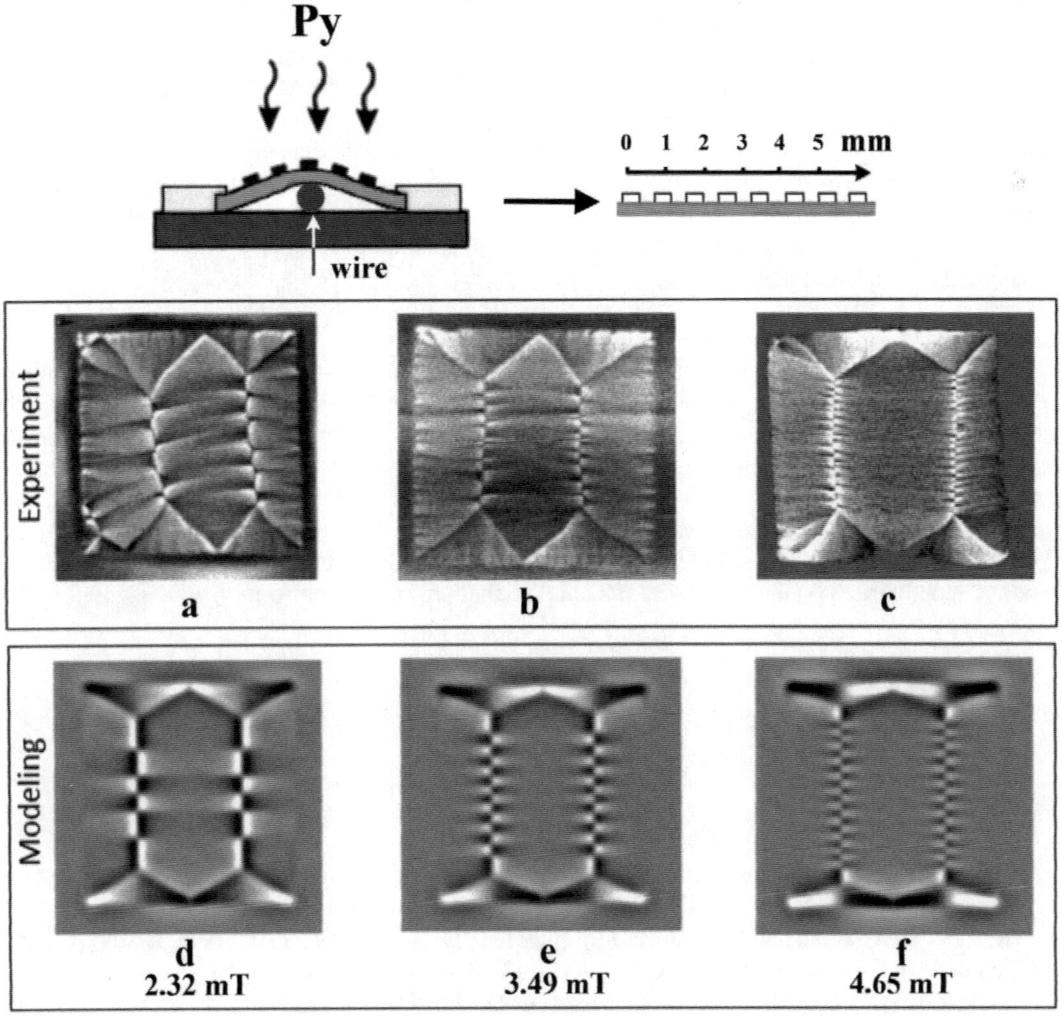

Figure 2. Scheme of the preparation of the wire-stressed sample (top), MFM scans of Py particles with dimensions of 25×25×0.03μm^3 on different substrate regions (bottom). Experimental: a) 0.1 mm from the substrate edge, b) 1.6 mm from the substrate edge, c) 2.6 mm from the substrate edge (sample center). Modelling: d, e, f. Corresponding effective anisotropy fields H_{eff} (2.32, 3.49 and 4.65 mT) for each particle used for the simulation.

Py microparticles with dimensions of 25×25×0.03 μm^3 were fabricated on the 0.30 mm thick glass surface. Using glass substrates and their nonuniform bending ability made it possible to fabricate samples with the elastic stress gradient increasing from the edge to the center of the substrate (analogously to [3], where the same technique was used for preparing mechanically stressed ferromagnetic films). The substrate was bent elastically in a holder by the 80 μm diameter wire placed under it prior to the Py deposition (figure 2). The particles were compressed uniaxially with the rate different for different places of the sample after straightening of the substrate.

MFM scans of stressed particles made it possible to establish that they have a seven-domain structure, which depends on the position of the particle on the substrate (figure 2a– c). Computer simulation was used to estimate the value of the induced magnetoelastic anisotropy in particles and its dependence on the location on the substrate. At the first stage, the magnetization distribution was calculated using the OOMMF program [7]. Topography scans of the particle obtained during MFM measurements were used in calculations. The effective anisotropy field H_{eff} was set as a variable parameter. At the second stage, the MFM scan of the particle was simulated by the "Virtual magnetic force microscope" program [8] using the magnetization distribution obtained with the OOMMF program. The parameter H_{eff} was varied until simulated and experimental MFM scans coincided.

The value of the effective anisotropy field of Py microparticles was estimated as a function of the mechanical stress by simulation their MFM images (figure 2). It was shown that the effective anisotropy field in wire-stressed samples increases in comparison with that in unstressed samples. Owing to the nonuniform distribution of the stress tensor the H_{eff} value of the particles increases from 2.32mT (the least compressed particles) on the edge of the sample to 4.65mT (the most compressed particles) in the sample center.

3. Conclusions

Using MFM measurements and computer simulation it was shown that the mechanical compression of the Py microparticles results in the increase in its effective anisotropy field. The redistribution of the particle magnetization is caused by the change of the H_{eff} value of the particle and depends on the stress tensor of the substrate in place where the particle is located. In general, when the substrate is stressed nonuniformly, MFM images of similar Py particles differ noticeably so it makes it possible to estimate the nonuniformity rate of stress induced in the substrate.

The mechanical compression of the permalloy microparticles leads to the noticeable decrease in the external magnetic field value necessary for the formation of the uniform magnetization in the compressed particle.

Acknowledgments

This work is supported in part by "Russian Foundation for Fundamental Research" (RFFI No. 15-02-02728).

References

[1] Morozov A I 2014 *Physics of the Solid State* **56** 865
[2] Barangi M and Mazamder P 2015 *IEEE Nanotechnology Magazine* **9**15
[3] Belyaev B A and Izotov A V 2007 *Physics of the Solid State* **49** 1731
[4] Bizyaev D A, Bukharaev A A, Kandrashkin Yu E, Mingalieva L V, Nurgazizov N I and Khanipov T F 2016 *Technical Physics Letters* **42** 1034
[5] Chuklanov A P *et al* 2016 *Journal of Physics* C **714**012006
[6] Tiercelin N, Dusch Y, Klimov A, Preobrazhensky V, Giordano S and Pernod P 2011 *J. Appl. Phys.* **109** 07D726
[7] Donahue M J and Porter D G Object oriented micromagnetic framework (OOMMF) (http://math.nist.gov/oommf/)
[8] Ovchinnikov D V and Bukharaev A A 2001 *Technical Physics* **46** 1014

Probing optical near-fields with photoreactive azo-polymers

M D Dmitrieva, A R Asadullina and M Kh Salakhov

Department of Optics and Nanophotonics, Institute of Physics, Kazan Federal University, Kremlevskaya st., 16, Kazan, 420008, Russian Federation.

E-mail: dmitruevam@gmail.com

Abstract. In this work, we demonstrate a method to visualize optical near-fields on photosensitive azo-polymer thin films with scanning near-field optical microscopy. A near-field intensity profile is determined in a basis of surface deformations of the azo-polymer thin film exposed to linearly polarized light.

1. Introduction

One of the most important issues in modern photonics and spectroscopy is the control and evaluation of the polarization state of optical near-fields. A control of the near-field polarization allows one to manipulate the light beyond the Abbe's diffraction limit. This is mainly used in non-volatile data storage and molecular switching [1, 2, 3, 4].

Besides, the near-field polarization plays a critical role in different optical techniques, such as scanning near-field microscopy (SNOM/NSOM) and tip-enhanced Raman spectroscopy (TERS) [5]. The control of near-field polarization enables to perform single-molecule sensitive experiments [6], to probe longitudinal and transverse optical anisotropy [7], and to determine the spatial orientation of molecules, and their symmetry. Also, it allows one to improve the spatial resolution in optical microscopy and spectroscopy [8]. The monitoring of the near-field enriches 3D polarized Raman microscopy [9, 10, 11] and near-field fluorescent microscopy [12].

The polarization of the light has a significant effect on the intensity distribution in the light beam. Due to the deformations of the polymer under the influence of incident light, we can observe the intensity profile of a polarized beam. This approach is valid both for far-fields and near-fields of the light.

In this paper, we demonstrate a method for visualization of the polarized optical near-fields on photosensitive azo-polymer thin films for the scanning near-field optical microscopy.

2. Experimental

In order to study the polarization of the near-field aperture probe, we realized a scheme of a scanning near-field microscope in transmission geometry based on a scanning probe microscope equipped with a confocal optical spectrometer NTEGRA Spectra (NT-MTD) (SNOM, figure 1). A commercial probe (NT-MTD) was taken as a test of the aperture of the probe. We applied photosensitive films a side-chain 4-amino-4-nitroazobenzene chromophore covalently attached to the polymer backbone of two epoxy-based oligomers containing hydroxyl groups (CFAO) Films were prepared by dissolving 7% wt CFAO compound in 1 mL cyclohexanone. A CFAO thin film was produced by spin coating 5 μl CFAO solution, in addition diluted by 10 times with cyclohexanone, onto an ITO/glass substrate at

3 000 rpm for 3 min, followed by annealing at the glass transition temperature of T_g = 130°C and 10 mbar for 1 h in order to remove solvent remains. We used this sample, due to its visible absorption spectrum with a maximal intensity peaked at 490 nm.

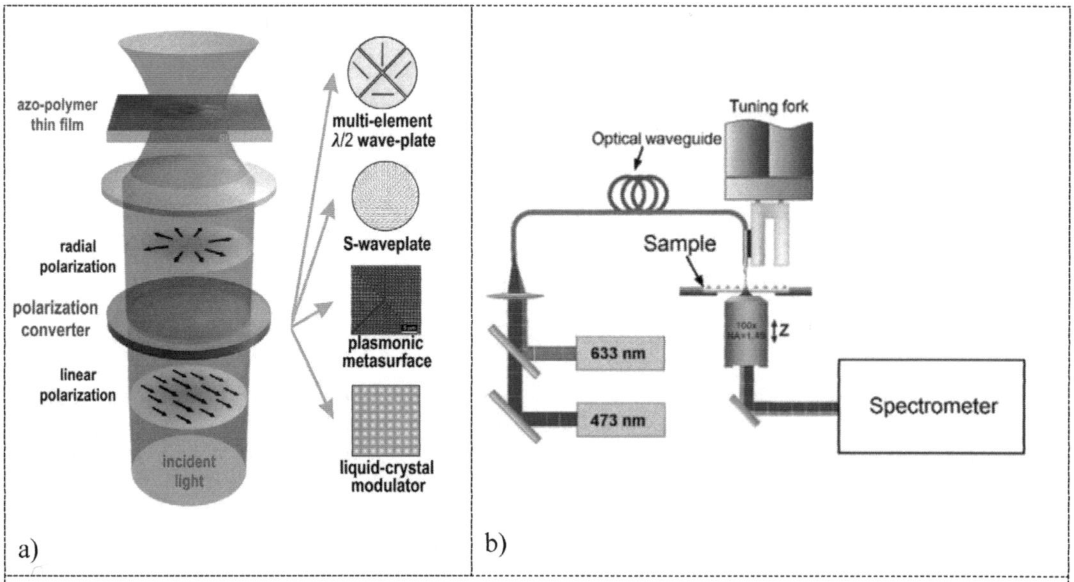

Figure 1. a) The mechanism of producing surface deformations by high-order Gaussian beams and b) the experimental setup of SNOM.

To control the spatial resolution of our setup, we used a standard optical grid (SN01, NT-MTD) for near-field measurements. Figure 2 represents lateral-force microscopy (A), confocal microscopy (B), and scanning near field microscopy (C) images of the optical grid SN01. On these data we can conclude, that near-field optical spatial resolution of this experimental setup amounts to 100 nm.

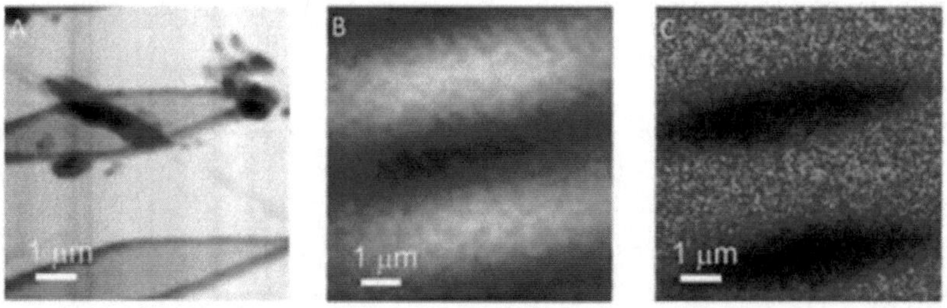

Figure 2. Lateral-force microscopy imaging (A), confocal (B) and near field (C) imaging of standard optical grid. Spatial resolution ~ 100nm.

We conducted a concomitant experiment on the deformation of the azo-polymers by incident light. In our experiment, we study photo-induced deformation of CFAO films produced by focused high-order Gaussian beams. A 633 nm He-Ne laser beam is focused into a spot of 200 nm in diameter using an 100x objective with numerical aperture NA=1.49. We use a half-wave S-waveplate to produce the radial- and azimuthal- polarized laser beams (ALTECHNA). Also, we did the experiment for linear polarized Gaussian beams. We used a hybrid SPM-equipped Raman spectrometer (NTEGRA SPECTRA, NT-MDT) was used to make topographic images of surface deformations. To produce the surface deformations on the CFAO thin film, the time exposure was 1 min.

IOP Conf. Series: Journal of Physics: Conf. Series **859** (2017) 012006 doi:10.1088/1742-6596/859/1/012006

3. Theoretical background

Azopolymers undergoes photoisomerisation under the incident light. The consequence of this phenomenon is motions of the chromophores. There are three types of motions in azopolymers: the first one is choromophores motions, which is caused by polarization of the incident light. As for linearly polarized light, the photoisomerisation is activated by the chromophore's transition dipole moment axis has a component parallel to the light polarization. The second type is the 'domain' motion. It requires the bounds between choromphores and the polymer matrix. The last type of motion appears on macroscopic scale, which reason is pressure gradient. [13, 14].

Bian et al first discovered the photo-induced deformation of azopolymer by means of focused Gaussian beams [15, 16, 17, 18]. In the polarizable azopolymer film an optical electric field $\boldsymbol{E}(\boldsymbol{r},t)$ induces a polarization $\boldsymbol{P}(\boldsymbol{r},t) = \epsilon_0\chi\boldsymbol{E}(\boldsymbol{r},t)$, where ϵ_0 is the permittivity of vacuum and χ is the medium susceptibility. The time averaged force density exerted on azochromophores in a small volume is represented by:

$$\boldsymbol{f}(\boldsymbol{r}) = \langle \boldsymbol{P}(\boldsymbol{r},t) \times \boldsymbol{\nabla}, \boldsymbol{E}(\boldsymbol{r},t)\rangle, \tag{1}$$

where $\boldsymbol{P}(\boldsymbol{r},t)$ is the film polarizability.

The field deforms azopolymers according to the law:

$$S(x,y,t) = \frac{1}{4}h\mu\epsilon_0\chi' \int_0^T \boldsymbol{\nabla}_s \times \boldsymbol{f}_{z=0}dt \tag{2}$$

where h is thickness of the polymer, μ is the proportionality factor that depends on the viscosity and form of polymeric chains, ϵ_0 is the permittivity in vacuum, χ' is real part of the susceptibility of the polymer, T is time exposure, $\boldsymbol{f}$ is force, that deforms polymer.

4. Results and discussions

In our work we applied this theoretical description of deformations [17, 18] and used it for our mathematical modeling of radial- and azimuthal-polarized Gaussian beams. Furthermore, mathematical modeling of radial- and azimuthal-polarized light was performed using the Jones matrix method. As a result, we gained calculation for linear, radial and azimutal polarizations for Gaussian beams.

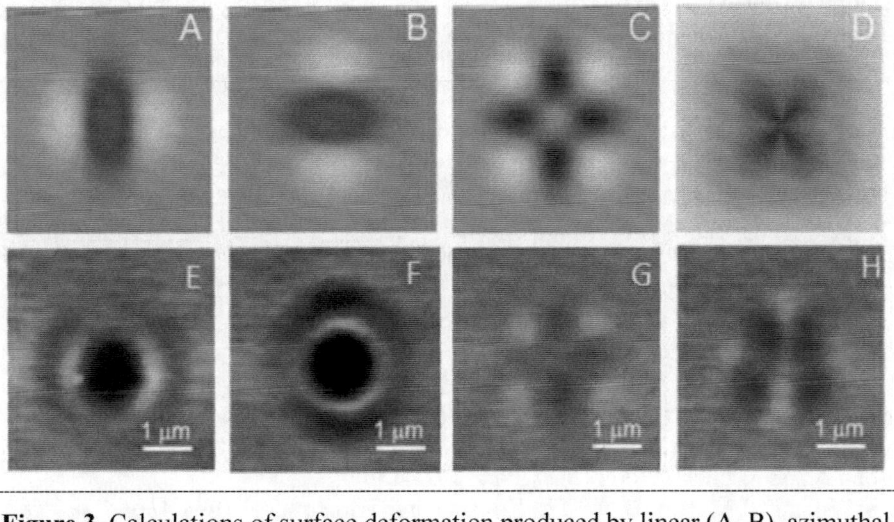

Figure 3. Calculations of surface deformation produced by linear (A, B), azimuthal (C) and radial (D) polarized Gaussian beams. AFM images of surface deformation produced by linear (E, F), azimuthal (G) and radial (H) polarized Gaussian beams.

The visualization of near and far fields has been achieved with photo-induced surface deformations. For linear, radial and azimuthal polarization, we obtain AFM images of surface deformations (Figure 3 (E), (F), (G), (H)). In figure 3 (E-H), the distribution corresponds to the dark areas of maximum light intensity, while the maximum surface deformations arise in the maxima gradient of the optical field.

By using the photo induced deformations method, we receive the near-field intensity profile for the aperture probe in scanning near-field microscopy (figure 4). A near-field intensity profile is determined on the basis of surface deformations of the azo-polymer thin film exposed to linearly polarized light. In figure 4, dark areas correspond to a maximum intensity of the optical near –field.

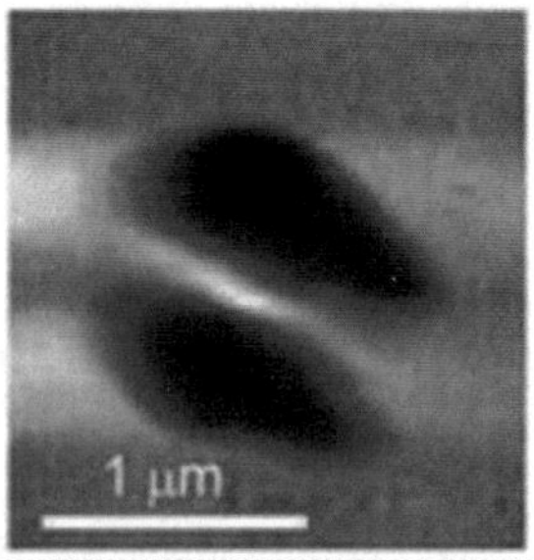

Figure 4. The near field distribution from SNOM probe on CFAO polymer.

5. Conclusion

We demonstrate a visualization technique for near field light in scanning near field microscopy by using photosensitive azo-polymer films. In summary, we conclude that the direction of linear polarization changes during the propagation in the optical fiber and localization process in SNOM tip.

References

[1] Meng X, Natansohn A, Barrett C and Rochon P 1996 *Macromolecules* **29** 946
[2] Ikeda T and Tsutsumi O 1995 *Science* **268** 1873
[3] Kharintsev S S, Fishman A I, Saikin S K and Kazarian S G 2016 *Nanoscale* **8** 19867
[4] Kharintsev S S, Fishman A I, Kazarian S G, Gabitov I R and Salakhov M Kh, 2014 *ACS Photonics* **1** 1025
[5] Kharintsev S S, Hoffman G G, Dorozhin P S, de With G and Loos J 2007 *Nanotechnology* **18** 315502
[6] Tauber T, Eilmann F and Hillenbrant R 2005 *Opt. Express* **13** 8893
[7] Sonntag M D, Klingsporn J M, Garribay L K, Roberts J M and Dieringer J A 2012 *Phys. Chem.* **116** 478
[8] Celebrano M, Kukura P, Renn A and Sandoghdar V 2011 *Nat. Photonics* **5** 95
[9] Florio G, Brundermann E, Yadavalli N S, Santer S and Havenith M 2014 *Soft Matter* **10** 1554
[10] Kharintsev S S, Fishman A I, Kazarian S G and Salakhov M Kh 2015 *Phys. Rev.* B **92** 115115-1
[11] Mino T, Satio Y and Verma P 2014 *ACS Nano* **8** 10187
[12] Vickery S A and Dunn R C 1999 Biophys. J. **76** 1812
[13] Nakajima K *et al* 1998 *Ultramicroscopy* **71** 257-62
[14] Mitsuoka Y *et al* 1998 *J. Appl. Phys.* **83** 3998
[15] Kharintsev S S, Chernykh E A , Fishman A I, Saikin S K, Alekseev A M and Salakhov M Kh 2017 *J. Phys. Chem.* **121** 3007
[16] Natasohn A and Rochon P 2002 *Chem. Rev.* **102** 4139
[17] Bian S, Williams J M, Kim D Y, Li L, Kumar J and Tripathy S K 1998 *Appl. Phys. Lett.***73** 1817
[18] Bian S, Williams J M, Kim D Y, Li L, Balasubramanian S and Kumar J 1999 *J. Appl. Phys.* **86** 4498

IOP Conf. Series: Journal of Physics: Conf. Series **859** (2017) 012007 doi:10.1088/1742-6596/859/1/012007

Suppression of transition dipole moments in radioactive atoms

R Kh Gainutdinov, A S Petrova and A A Mutygullina

Kazan Federal University, 18 Kremlevskaya str., 420008 Kazan, Russian Federation

E-mail: Renat.Gainutdinov@kpfu.ru

Abstract. Optical emission from radioactive atoms is investigated. It is shown that the radioactivity of the central nucleus results in the suppresion of the optical transition dipole moment in the corresponding atom. The situations are regarded when this suppression is significant and can be experimentally observed.

1. Introduction

Optical research techniques play an important role for nuclear physics [1]. Measuring isotopic shift and hyperfine splitting in atomic spectra provides us information on nuclear mass, shape and spins [2,3]. The optical approach is especially important for short-lived nuclei, where other techniques may appear not applicable. However, in the case of short-lived nuclei the ratio of the mean time of an optical transition and the nuclear half-life comes into play. Obviously, the lifetime of a radioactive nucleus should be larger or at least comparable to the time required for the optical transition to occur. In this paper we will show, that in the case when these times are of the same order, nuclear radioactivity manifests itself in the optical spectrum, giving rise to the suppresion of the optical transition dipole moment, and as a consequence to the decrease of spontaneous emission probability.

2. Natural broadening

It is well known that the natural broadening of a spectral line emitted on the transition from an initial level $|i\rangle$ to a final level $|f\rangle$ is determined by the Breit-Wigner contour:

$$\frac{dW}{d\omega} = \frac{2\pi}{(\omega - \omega_{fi})^2 + \Gamma_{fi}^2/4}, \tag{1}$$

where ω is the photon frequency, ω_{fi} is the atomic transition frequency, Γ_i (Γ_f) is the width of the initial (final) level, and $\Gamma_{fi} = \Gamma_f + \Gamma_i$ is the spectral line width.

An atom represents a bound system consisting of a nucleus and electrons and should be described by a multiparticle wave function. Correspondingly, the width of an atomic level Γ_n ($n = i, f$) is affected both by the nuclear and electronic decay processes and in the simplest case can be represented as the superposition:

$$\Gamma_n = \Gamma_n^{(e)} + \Gamma_0^{(N)}, \tag{2}$$

where $\Gamma_n^{(e)}$ is the purely electromagnetic decay width and $\Gamma_0^{(N)}$ is the nuclear decay width. Here we suppose the nuclear decay processes to be not affected by the state of the electronic shells, i.e. $\Gamma_0^{(N)}$ is one and the same for any electronic state $|n\rangle$. This suggestion is justifiable, because the nuclear interaction is several orders more intensive than the electromagnetic one. From this it follows, that the spectral line width Γ_{fi} is generally described by the following expression:

$$\Gamma_{fi} = \Gamma_{fi}^{(e)} + 2\Gamma_0^{(N)}, \tag{3}$$

where $\Gamma_{fi}^{(e)} = \Gamma_f^{(e)} + \Gamma_i^{(e)}$ is the width of the electronic transition. Usually, $\Gamma_{fi}^{(e)} \gg \Gamma_N^{(0)}$, i.e. the nuclear lifetime $\tau_N = \hbar/\Gamma_0^{(N)}$ is much greater than the mean time of the electronic transition $\tau_{fi}^{(e)} = \hbar/\Gamma_{fi}^{(e)}$. This ensures that the atom emits the photon till its nucleus decays. The spectral line width in this case is determined by the electronic decay width only: $\Gamma_{fi} = \Gamma_{fi}^{(e)}$. In contrast, when $\Gamma_{fi}^{(e)} \ll \Gamma_N^{(0)}$, i.e. $\tau_{fi}^{(e)} \gg \tau_N$, the nucleus decays before the atom emits and no optical spectrum can be observed.

However, for some atoms created under laboratory conditions the nuclear lifetime coincides in order with the mean time of an optical transition [4,5]. For such atoms the widths in expression (3) are comparable and contribute equally into the total spectral line width. Thus, the natural line broadening of such atoms will be affected by the nuclear radioactivity as well.

3. Generalized Quantum Dynamics

While the electromagnetic interaction is well understood and theoretically described, the nuclear interaction Hamiltonian has not been found yet. Instaed, the effective field theory (EFT) was proposed by Weinberg for the description of nucleon-nucleon (NN) interactions, which implies no potentials at all [6]. At the same time, it has been shown that the Schrödinger equation is not the most general dynamical equation consistent with the current concepts of quantum physics, and a more general equation of motion was derived by Gainutdinov in [7]. Being equivalent to the Schrödinger equation in the case of instantaneous interactions, this equation permits the generalization to the case where dynamics is governed by a nonlocal-in-time interaction. It was successfully applied for solving different problems in atomic [8] and nuclear physics [9–11], as well as in quantum optics [12]. In Ref. [9] it has been shown that the requirement that the NN T-matrix satisfies this equation and the symmetries of quantum chromodynamics, determines the low energy NN operator up to a constant. The interaction operator defined in this way is nonlocal in time. Thus, the low energy NN dynamics is governed by the Gainutdinov equation with the nonlocal-in-time interaction operator. The NN T-matrix obtained by solving this equation was shown to coincide with that derived in EFT of nuclear forces [13]. In this work we use this equation, because its form does not depend on the specific feature of the interaction (the Scrödinger equation contains the interaction Hamiltonian). All the needed dynamical information is contained in the boundary condition for this equation. This feature of the equation allows one to draw general conclusions about the decay of such complex systems as nuclei.

The Gainutdinov equation has been derived as a direct consequence of the basic postulate of the Feynman approach expressing the superposition law for quantum probability amplitudes and the basic postulates of the canonical approach manifesting the ingenious Dirac guess that the physical logic of microworld is adequate to the mathematical logic of Hilbert spaces. In particular, from the Feynman postulate it follows that $\langle\Psi_2|U(t_2,t_1)|\Psi_1\rangle$, being the probability amplitude of finding the quantum system in the state $|\Psi_2\rangle$ at time t_2 if at time t_1 it was in the state $|\Psi_1\rangle$, can be represented as a sum of probability amplitudes for each of the alternative ways in which this event can happen. These alternatives can be divided in different classes. For

IOP Conf. Series: Journal of Physics: Conf. Series **859** (2017) 012007 doi:10.1088/1742-6596/859/1/012007

example, subprocesses with definite instants of the beginning and end of the interaction in the system can be chosen as such alternatives. In this case the evolution operator can be represented in the form

$$U(t_2,t_1) = \mathbf{1} + \int_{t_1}^{t_2} dt' \int_{t_1}^{t'} dt\ \tilde{S}(t',t), \tag{4}$$

where $\tilde{S}(t',t)$ is an operator whose matrix elements $\langle\Psi_2|\tilde{S}(t',t)|\Psi_1\rangle$ are the probability amplitudes for the alternative ways in which the interaction in the system begins at time t and ends at time t'. For the evolution operator to be unitary for any t_2 and t_1, the operator $\tilde{S}(t',t)$ must satisfy the equation [7]:

$$(t'-t)\tilde{S}(t',t) = \int_t^{t'} dt_4 \int_t^{t_4} dt_3 (t_4 - t_3)\tilde{S}(t',t_4)\tilde{S}(t_3,t). \tag{5}$$

This equation allows one to obtain the operator $\tilde{S}(t',t)$ for any t and t' and hence the dynamics of the system, if $\tilde{S}(t_4',t_3')$ corresponding to infinitesimal duration times $\tau = t_4' - t_3'$ of interaction are known. Most of the contribution to the evolution operator in the limit $t_2 \to t_1$ comes from the processes associated with the fundamental interaction in the system encoded in the interaction operator $H_{\text{int}}(t',t)$. Equation (5) with the boundary condition $\tilde{S}(t',t)\lim_{t'\to t} H_{\text{int}}(t',t)$ determines the dynamics of the system. The evolution operator $U(t,0)$ in the Schrödinger picture can be presented in the form:

$$U(t,0) = \frac{i}{2\pi}\int dx \exp(-izt) G(z), \tag{6}$$

where $z = x + iy$, $G(z) = G_0(z) + G_0(z)T(z)G_0(z)$, $G_0(z) = (z - H_0)^{-1}$ is the free Green operator (H_0 is the free Hamiltonian) and $T(z) = i\int_0^\infty d\tau \exp^{(-iz\tau)} \exp(-iH_0t_2)\tilde{S}(t_2,t_1)\exp(iH_0t_1)$. The contribution to the Green operator $G(z)$, which comes from the processes associated with the self-interaction of particles, has the same structure as the free Green operator $G_0(z)$. For this reason it is natural to replace $G_0(z)$ by the propagator $\tilde{G}_0(z)$, which describes the evolution of particles interacting only with vacuum and hence has the structure $\tilde{G}_0(z) = (z - H_0 - C(z))^{-1}$, where the operator $C(z)$ has the same eigenvectors as H_0 ($C(z)|n\rangle = C_n(z)|n\rangle$, $H_0|n\rangle = E_n|n\rangle$). Correspondingly, the operator $T(z)$ should be replaced by the operator $M(z)$, which describes the evolution of particles interacting not only with vacuum. These operators are related as follows: $G(z) = G_0(z) + G_0(z)T(z)G_0(z) = \tilde{G}_0(z) + \tilde{G}_0(z)M(z)\tilde{G}_0(z)$. Making use of the equation

$$\frac{dG(z)}{dz} = -G^2(z), \tag{7}$$

being one of the forms of equation (5) in the energy representation [14–18], one can get the following equations for the operators $M(z)$ and $C(z)$:

$$\begin{aligned} \frac{dM(z)}{dz} &= -(1-P)M(z)\tilde{G}_0(z)\tilde{G}_0(z)M(z) \\ &- M(z)\tilde{G}_0(z)\frac{dC(z)}{dz} - \frac{dC(z)}{dz}\tilde{G}_0(z)M(z), \end{aligned} \tag{8}$$

$$\frac{dC_n(z)}{dz} = -\langle n|M(z)(\tilde{G}_0(z))^2 M(z)|n\rangle, \tag{9}$$

where P is a projection operator: $P\hat{A} = \sum_n |n\rangle\langle n|a_n$ with $a_n = \langle n|\hat{A}|n\rangle$. The boundary conditions for these equations are $C(z)\lim_{|z|\to\infty} PB(z)$ and $M(z)\lim_{|z|\to\infty}(1-P)B(z)$, where $B(z) = i\int_0^\infty e^{-iz\tau}e^{-iH_0t_2}H_{\text{int}}(t_2,t_1)e^{iH_0t_1}$.

In the case of a radioactive atom, the interaction operator $B(z)$ is a sum of the operator describing the strong interaction of nucleons $B^{(N)}(z)$ and the Hamiltonian H_I describing the week interaction of the atomic electrons with the electromagnetic field $B(z) = B^{(N)}(z) + H_I$. Solving equations (8) and (9) by expanding the solution in powers of H_I and restricting ourselves to the second order term, we get

$$M(z) = M^{(N)}(z)+H_I+M^{(N)}(z)\tilde{G}_0(z)H_I+H_I\tilde{G}_0(z)M^{(N)}(z)+H_I\tilde{G}_0(z)M^{(N)}(z)\tilde{G}_0(z)H_I, \quad (10)$$

$$C_n(z) = C_n^{(e)} + C_n^{(N)}, \quad (11)$$

where $C_n^{(e)}(z)$ and $C_n^{(N)}(z)$ are the solutions of the equations $dC_n^{(e)}(z)/dz = -\langle n|H_I(\tilde{G}_0(z))^2 H_I|n\rangle$ and $dC_n^{(N)}(z)/dz = -\langle n|M^{(N)}(z)(\tilde{G}_0(z))^2 M^{(N)}(z)|n\rangle$ respectively with corresponding boundary conditions. $M^{(N)}$ is the solution of equation (8) with the boundary condition $M(z)\lim_{|z|\to\infty}(1-P)B^{(N)}(z)$.

Generally, the self energy function $C_n(z)$ can be represented by its Taylor expansion:

$$C_n(z) = C_n(z = E_n^{(0)}) + \frac{dC_n(z)}{dz}|_{z=E_n^{(0)}} + ..., \quad (12)$$

where $E_n^{(0)}$ is the bare energy of the state $|n\rangle$. It is well known that the first term in this expansion represents the energy shift $\Delta E_n^{(0)}$ caused by the vacuum polarization and the decay width Γ_n:

$$C_n(z = E_n^{(0)}) = \Delta E_n^{(0)} - i\Gamma_n.$$

In what follows we will assume that the energy shifts $\Delta E_n^{(0)}$ are included into the energies of the states, i.e. that the renormalized energy values are used.

As for the derivative

$$\eta = \frac{dC_n(z)}{dz}|_{z=E_n^{(0)}},$$

it includes the contributions from the off-shell virtual transitions associated with vacuum fluctuations $\Sigma_n(z)$ and from the real on-shell transitions giving rise to the system dissipation $C_n^{\text{on}}(z)$: $\eta = \Sigma_n(z) + C_n^{\text{on}}(z)$. The second term in this expression is imaginary and represents a small correction to the decay width Γ_n, which can be neglected in virtue of the exponentiality of the decay law. The off-shell function characterizes the strength of quantum fluctuations, manifesting themselves in virtual transitions of the quantum system into all possible decay states. The fact that in the problem under study the decay widths of the nuclear and electronic states are of the same order of magnitude does not mean that the same is true for the nuclear and electronic off-shell functions $\Sigma_n^{(N)}(z)$ and $\Sigma_n^{(e)}(z)$. Indeed, despite the fact that the interaction of nucleons is strong, the nucleon decay widths that are determined by the on-shell transitions are relatively small. But in virtual transitions the strength of NN interaction may manifest itself explicitly. In contrast, the quantum fluctuations induced by the interaction of electrons with the electromagnetic field are relatively week ($\Sigma_n^{(e)}$ are of the order of $\alpha^4 m_e$, where $\alpha = 1/137$ is the fine structure constant and m_e is the electronic rest mass) and can be neglected. Thus, the self energy function of an atom can be represented in the form

$$C_n(z) = \Sigma_n^{(N)}(z) - i\Gamma_0^{(N)}/2 - i\Gamma_n^{(e)}/2. \quad (13)$$

With (13), the free green operator of the radioactive atom in the state $|n\rangle$ has the following form:

$$G(z) = \frac{1}{z - E_N - \varepsilon_n - H_I - B^{(N)}(z) - \Sigma_n^{(N)}(z) + i\Gamma_0^{(N)}/2 + i\Gamma_n^{(e)}/2},$$

where E_N is the renormalized rest energy of the nucleus and ε_n is the electronic energy in the state $|n\rangle$.

4. Suppression of atomic transition dipole moment

The probability A_{fi} that a photon will be emitted, when the atom transits from the state $|i\rangle$ to the state $|f\rangle$ is proportinal to the cube of the photon frequency ω and the absolute square of the matrix element of transition dipole moment operator d_{fi}:

$$A_{fi} \sim \omega^3 |d_{fi}|^2.$$

The matrix element of the transition dipole moment operator is defined on the basis of the electronic wave functions φ_f and φ_i in the initial and in the final states correspondingly:

$$d_{fi} = e \int \varphi_f^*(\mathbf{r})\mathbf{r}\varphi_i(\mathbf{r})d\mathbf{r}. \tag{14}$$

These functions are usually found as solutions of the corresponding Schrödinger (Dirac) equation with the constant central potential.

However, due to quantum fluctuations the unstable nucleus spends a part of time in its bare state $|\Psi_n^0\rangle$, while another part of time it exists in the virtual state of its decay products $|\Psi_{\text{virt}}\rangle$. Thus, the state vector $|\Psi_n\rangle$ of the unstable nucleus can be generally represented as a superposition of two vectors:

$$|\Psi_n\rangle = \mathcal{Z}|\Psi_n^0\rangle + \chi|\Psi_{\text{virt}}\rangle,$$

where $\mathcal{Z} = (1 - \eta)^{-1}$ determines the probability to find the nucleus in the bare state, while $\chi = \eta/(\eta - 1)$ is the probability to find the nucleus in the state of its decay products. This means, that the atomic electrons spend not the whole time in the field of the bare nucleus potential, when they can be described by the wave functions φ_f and φ_i. Actually, during a certain time they exist in the field of the nuclear decay products. Owing to this, the contribution from φ_f and φ_i to (14) decreases and thus the matrix element of the dipole moment operator is suppressed. It can be shown [19], that the probability for a photon to be emitted with a certain energy ω is defined by the following formula:

$$\frac{dW}{d\omega} \sim \omega \frac{\Gamma}{(\omega - \omega_{fi})^2 + \Gamma^2/4}, \tag{15}$$

with the width

$$\Gamma = 2\Gamma^{(N)} + \mathcal{Z}^2\Gamma_{fi}^{(e)}. \tag{16}$$

According to the definition, $\mathcal{Z} < 1$, and hence the contribution from the electromagnetic decay width is suppressed.

The coefficients $\mathcal{Z}$ and χ determine the ratio of the time spent by the nucleus in the bare state to the time spent in the state of its decay products. The values of these parameters change from one isotope to another, as well as depend on the nuclear decay type. Usually $\mathcal{Z} \sim 1$, but in some situations this relation may change. One interesting example for this is super-allowed alpha decay observed for certain light nuclei in the vicinity of ^{100}Sn isotope. It is supposed that the possibility of alpha decay in this region is caused by the fact that protons and neutrons occupy identical orbitals leading to a significant increase of the alpha particle preformation probability [20]. This means, that the part of time spent by the nucleus in the state of its decay products is essential. The lightest experimentally observed alpha emitter is the isotope of tellurium ^{105}Te. The measured reduced width of this isotope is $\delta_r \sim 70keV$ [4, 5]. It can be shown that the corresponding strength of quantum fluctuations should have the order of 0.4: $\eta \sim 0.4$. Additionally, the half-life of ^{105}Te is measured to be $\sim$ 700ns, that is the widths $\Gamma^{(N)}$ and $\Gamma_{fi}^{(e)}$ are comparable for certain optical transitions in the atom of this isotope. Thus, according to formulae (15) and (16) the nuclear radioactivity will manifest itself in the optical spectrum, giving rise to the change of natural line broadening.

5. Conclusion

We have considered the influence of nuclear radioactivity on the optical spectrum of corresponding atoms. We have derived a general formula for the spectral line profile, emitted from an atom with a radioactive nucleus, which contains the nuclear decay width as well. We have shown that for certain nuclei quantum fluctuation processes result in the suppression of the atomic transition dipole moment and as a consequence to the change of spectral line natural width. Physically, the reason for this is the fact, that such nuclei spend just a part of time in a bare state, while another part of time they exist in the state of their decay products, and as a consequence the atomic electrons experience the bare nucleus potential not the whole time. This phenomenon can be important for the creation of new optical method for the measurement of nuclear half-life from the optical line width.

Acknowledgments

This work was funded by the subsidy allocated to Kazan Federal University for the state assignment in the sphere of scientific activities.

References

[1] Campbell P, Moore I D and Pearson M R 2016 *Prog. Part. Nucl. Phys.* **86** 127
[2] Cheal B and Flanagan K T 2010 *J. Phys.* G **37** 11
[3] Cheal B, Cocolios T E and Fritzsche S 2012 *Phys. Rev.* A **86** 4
[4] Seweryniak D *et al* 2006 *Phys. Rev.* C **73** 6
[5] Liddick S N *et al* 2006 *Phys. Rev. Lett.* **97** 8
[6] Weinberg S 1991 *Nucl. Phys.* **B363** 3
[7] Gainutdinov R Kh 1999 *J. Phys.* A **32** 30
[8] Gainutdinov R Kh, Mutygullina A A and Scheid W 2002 Phys. Lett. A **306** 1
[9] Gainutdinov R Kh and Mutygullina A A 2002 *Phys. Rev.* C **66** 1
[10] Gainutdinov R Kh and Moutygoullina A A 1997 *Physics of Atomic Nuclei* **60** 841-847
[11] Gainutdinov R Kh and Mutygullina A A 2002 *Physics of Atomic Nuclei* **65** 1421-1430
[12] Gainutdinov R Kh and Mutygullina A A 2009 *Physica Scripta* T **135** 014012
[13] Van Kolk U 1999 *Prog. Part. Nucl. Phys.* **43** 409
[14] Gainutdinov R Kh 1989 *J. Phys.* A **22** 269-286
[15] Gainutdinov R Kh 1983 *Sov. J. Nucl. Phys.* **37** 277-282
[16] Gainutdinov R Kh 1987 *Sov. J. Nucl. Phys.* **46** 743-749
[17] Gainutdinov R Kh 1987 *Sov. J. Nucl. Phys.* **53** 885-892
[18] Gainutdinov R Kh 1995 *Zh. Eksp. Teor. Fiz.* **108** 1600-1613
[19] Gainutdinov R Kh, Mutygullina A A and Petrova A S 2016 *J. Phys.: Conf. Series* **714** 012008
[20] Guo Sh, Bao X, Gao Y, Li J and Zhang H 2015 *Nucl. Phys.* A **903** 1.

IOP Conf. Series: Journal of Physics: Conf. Series **859** (2017) 012008 doi:10.1088/1742-6596/859/1/012008

Particle swarm optimization of broadband field enhancement with a grating-assisted plasmonic taper nanoantenna

A R Gazizov[1,2], **M Zohrabi**[1] **and M Kh Salakhov**[1,2]

[1] Kazan Federal University, 18 Kremlevskaya str., 420008 Kazan, Russian Federation
[2] Academy of Science of the Republic Tatarstan, 20 Bauman str., 420111 Kazan, Russian Federation

E-mail: Almaz.Gazizov@kpfu.ru

Abstract. This work is dedicated to the improvement of the near-field enhancement beneath the gold and silver tip apex due to plasmons excitation on a sub-wavelength grating engraved on the tip lateral surface. To study conditions of the maximal enhancement we have performed PSO-based optimization of intensity in search space of two parameters for gold and silver tip with different cone angles. Parameters of search space are period of the grating and its position in respect to the apex. The grating-assisted tip is illuminated with the incident light with wavelengths of 400 to 1000 nm in our model. All the simulations of electromagnetic waves scattering on the nanoantenna are based on the finite difference time domain method.

1. Introduction

In tip-enhanced optical microscopy and spectroscopy, taper metallic probes are commonly used to convert the freely propagating field to localized modes and vice versa [1]. They permit significant localization of incident light at a subwavelength scale and light-matter interaction enhancement on the sample [2]. Such optical antennas have applications in many fields of nanoscience [3-5]. Plasmonic nanoantennas are the one type of nanoantennas, in which the field-enhancement is provided by surface plasmon resonances [6]. For instance, in TERS epi-illumination configuration, the sample is under direct external laser illumination in the presence of a metallic tip near the sample. It causes the plasmon excitation at the metallic tip and the field enhancement, but at the same time a background signal from a diffraction-limited area appears. To achieve background suppression a grating-assisted plasmonic nanoantenna has been suggested [7]. Adiabatic nanofocusing is a suitable way for transforming the optical signal of the far field in the near field of the localized light source [8]. It has a huge potential to perform background-free nanoscale chemical spectroscopy. The investigation of the adiabatic focusing by grating-patterned optical antennas can be found in Refs. [9-11]. It is reasonable to use the nanoantennas with optimized grating parameters for the effective surface plasmon excitation.

In this work we aim to find optimal value of grating parameters for broadband external source to effective plasmon excitation and propagation. We considered gold and silver probes with different cone angles and curvature radii of 10 nm at the side-illumination configuration and performed the particle swarm optimization of electric field intensity at the tip apex. A global best search was performed for a grating period and its position in respect to the tip apex. To find the intensity near the tip apex the finite-difference time-domain (FDTD) simulation was used.

IOP Conf. Series: Journal of Physics: Conf. Series **859** (2017) 012008 doi:10.1088/1742-6596/859/1/012008

2. Results and discussion

Consider a taper metallic optical probe oriented along Z-axis with a grating on its surface (figure 1). A curvature radius ρ is 10 nm, cone angle α vary from 5 to 30 degrees. A probe material can be either gold or silver. The two parameters for particle swarm optimization are the period of grating (a) and spacing of the grating from the tip apex along the axis of the probe (b). They form two-parametric search space. The grating of the nanoantenna is illuminated by plane wave propagating along X-axis with a broadband spectrum with wavelengths from 400 to 1000 nm, so the angle of incidence is $\varphi_i = 90° - \alpha/2$.

One can express the complex enhancement factor of the field as follows:

$$\alpha_{\mathrm{eff},||} = 2\pi\varepsilon_0\rho^3 f_e \quad (1)$$

Expression (4) for the longitudinal polarizability of the probe occurs from the condition that the field on the surface of the edge is numerically equal to the field that is specified $f_e E_0$ [12]. Using the formalism of the Green's function one can obtain the expression of the resultant field [12-14]:

$$\mathbf{E}(\mathbf{r},\omega) = \mathbf{E}_0(\mathbf{r},\omega) + \frac{\omega^2}{\varepsilon_0 c^2}\ddot{\mathbf{G}}_0(\mathbf{r},\mathbf{r_0};\omega)\ddot{\alpha}_{\mathrm{eff}}\mathbf{E_0}(\mathbf{r_0},\omega) \quad (2)$$

To find the electric field $\mathbf{E}(\mathbf{r},\omega)$ we use the finite-difference time-domain simulation. FDTD method is a successful application in an extremely wide range of tasks, such as light scattering from metal objects and dielectrics, antennas, and microstrip circuits [15]. We also use an algorithm of particle swarm optimization (PSO). PSO algorithm is the one of the most powerful methods for solving the non-smooth global optimization problems [16]. We used the PSO algorithm with global best described in Ref. [17] and in our previous work [18] for optimization of the field intensity beneath the tip apex.

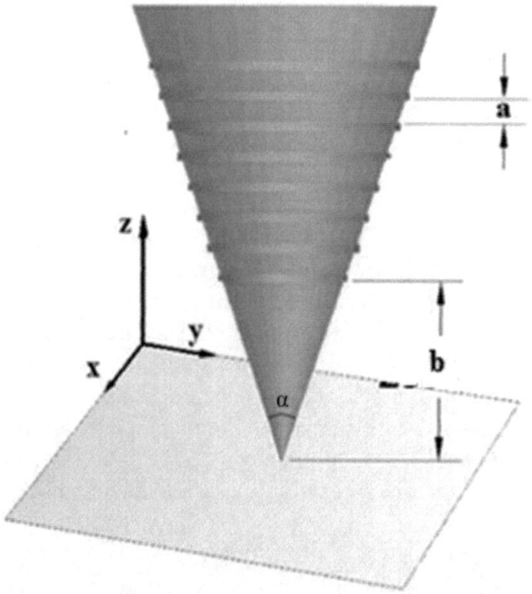

Figure 1. Schematic of a grating-assisted optical antenna.

The objective function is a sum of squared differences between the enhancements of the intensity near the tip apex for a given particle of the swarm and for other particles.

We used PSO algorithm and FDTD-simulation for each material of the probe (gold, silver) and for each cone angle (for 5, 10, 20, 30 degrees) separately. In each case we obtain intensity distribution beneath the tip apex in XY-plane and a power spectrum. For the numerical simulation the dielectric function of gold was taken from Ref. [18]

Figure 2 shows the intensity distributions around the tip in the XY-plane and spectra of power flux through XY-plane for gold and silver nanoantenna with the optimal parameters. For gold antenna period a=245 nm, b=1500 nm, cone angle α=20°; for silver those parameters are a=250 nm, b=1000 nm, α=5° respectively.

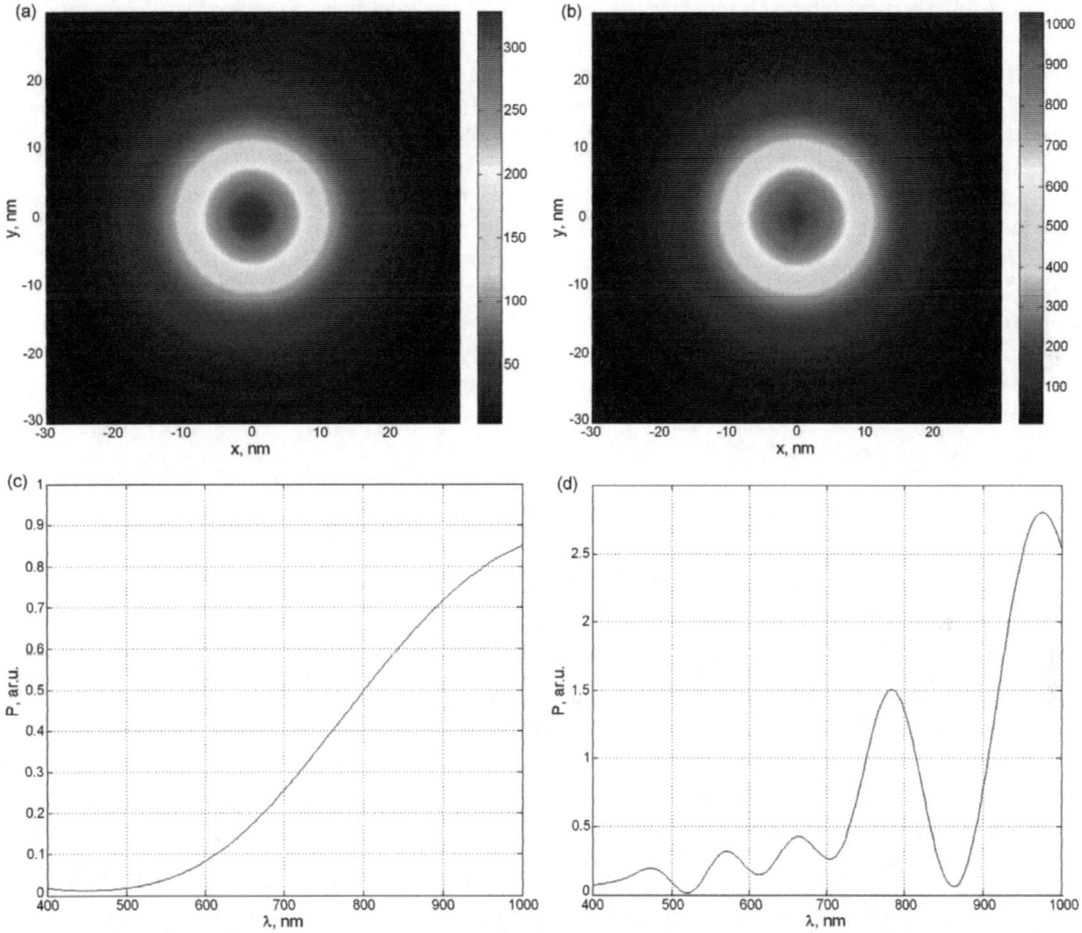

Figure 2. a) intensity distribution in the XY-plane of optimized gold tip, b) intensity distribution in the XY-plane of optimized silver tip, c) power flux through XY-plane spectrum for optimized gold tip, d) power flux through XY-plane spectrum for optimized silver tip.

In a case of illumination of the grating by broadband radiation with wavelengths of 400 to 1000 nm, we obtain enhancement of a total intensity of radiation beneath the tip apex. As we can see in figure 2 the absolute value of the enhancement factor $|f_e|$ reaches ~18 for gold tip and ~31 for silver tip in that spectrum region. For instance, for taper nanoantennas with the same cone angles but without grating we got enhancement of intensity by a factor 4.5 for gold tip and 5.5 for silver tip with direct illumination of the tip.

IOP Conf. Series: Journal of Physics: Conf. Series **859** (2017) 012008 doi:10.1088/1742-6596/859/1/012008

Frequency selectivity of the light-grating coupling occurs due to a phase matching of surface plasmon excitation and localized plasmon resonances of the considered tip. Those resonances depend on the curvature radius, the cone angle and the tip material, while the phase matching depends on the angle of incidence of the light on the grating, and the period of grating. The optimal location of the grating in respect to the apex exists, because there are two competing processes of plasmon enhancement and extinction. During propagation towards the apex, the surface wave undergoes losses within the probe, radiation losses, and enhancement due to mode compression. The cone angle affects radiative losses and mode compression, while grating position affects mode compression and losses in probe material. Thus, both period of the grating and its position have been optimized.

3. Conclusion

The results show that the grating-assisted gold and silver tip with optimized parameters can effectively compress the electric field and localize it within the extent of ~10 nm. The absolute value of the enhancement factor reaches ~18 for gold tip and ~31 for silver tip. In the case of side illumination one can expect a significant suppression of the background signal. Numerical simulations have shown resonance shifts when the grating period changes.

Acknowledgments
This work was funded by the subsidy allocated to Kazan Federal University for the state assignment in the sphere of scientific activities. This work was financially supported by the Russian Foundation for Basic Research (No. 15-42-02339). This work was done using equipment of Federal Center of Shared Equipment of Kazan Federal University.

References

[1] Novotny L and van Hulst N 2011 *Nature Photonics* **5** 83-90
[2] Maier S A and Atwater H A 2005 *J. Appl. Phys.* **98** 011101
[3] Georgi C and Hartschuh A 2010 *Appl. Phys. Lett.* **97** 143117
[4] Atwater H A and Polman A 2010 *Nature Mater.* **9** 205-213
[5] Bardhan R et al 2011 *Acc. Chem. Res.* **44** 936-946
[6] Novotny L 2011 *Physics today* **64** 47-52
[7] Berweger S, Atkin J M, Olmon R L and Raschke M B 2010 *J. Phys. Chem. Lett.* **1** 3427-3432
[8] Stockman M I 2004 *Phys. Rev. Lett.* **93** 137404
[9] Ropers C, Neacsu C C, Raschke M.B, Albrecht M, Lienau C and Elsaesser T 2008 *Japanese Journal of Applied Physics* **47** 6051-6054
[10] Ropers C, Neacsu C C, Elsaesser T, Albrecht M, Raschke M B and Lienau C 2007 *Nano Lett.* **7** 2784-2788
[11] Giugni A, Torre B, Tomal A, Francardi M, Malerbal M, Alabastri A, Proietti Zaccarial R, Stockman M I and Di Fabrizio E 2013 *Nature Nanotechnology* **8** 845-852
[16] Novotny L and Stranick S J 2006 *Annu. Rev. Phys. Chem.* **57** 303-331
[17] Maximiano R V, Beams R, Novotny L, Jorio A and Cançado L G 2012 *Phys. Rev.* B **85** 235434
[18] Novotny L and Hecht B 2006 *Principles of Nano-optics* (Cambridge: Cambridge University Press)
[12] Salski B, Celuch M and Gwarek W 2010 *IEEE Microwave Magazine* **11** 50-59
[13] Kennedy J and Eberhart R C 1995 *Proceedings of IEEE International conference on Neural Networks* **4** 1942-1948
[14] Das S, Abraham A and Konar A 2008 *Pattern Recognition Letters* **29** 688-699
[15] Gazizov A R, Zohrabi M, Kharintsev S S and Salakhov M Kh 2016 *J. Phys.: Conf. Ser.* **714** 012010
[19] Johnson P B and Christy R W 1972 *Phys. Rev.* B **6** 4370-4379

IOP Conf. Series: Journal of Physics: Conf. Series **859** (2017) 012009 doi:10.1088/1742-6596/859/1/012009

FTIR spectroscopy of conformational probes introduced into binary blends of polymers

D I Kamalova, L R Abdrazakova and S N Naumova

Kazan Federal University, Kazan 420008, Russian Federation

E-mail: dina.kamalova@kpfu.ru

Abstract. In this paper we considered the behaviour of the conformational probes introduced in polymer blends. The temperatures of the secondary relaxation transitions of the binary blends of polyvinyl butyral and polyethylene glycol were determined by FTIR spectra. The assignment of the secondary relaxation transitions to the types of local molecular mobility in the blend of polymers was carried out.

1. Introduction

Many physical and chemical properties of polymeric materials are due to local molecular dynamics of polymeric chains. One important characteristic of molecular mobility in polymers is the secondary relaxation transitions that appear below the glass transition temperature. They correspond to the "unfreezing" of the internal rotation of side groups and other fragments (relaxators) of the polymer chain when temperature increases. The method of FTIR spectroscopy of conformational probes allows determination of the temperature of relaxation transitions and assigning them to specific relaxators, the swing isomer mobility of which facilitates the diffusion of separated gases in polymer membranes, protective layers and coatings [1].

The blending of polymers is one of the methods of modification of properties of polymeric materials. This method allows one to create unique micro- and nanoscale structure that are not available in the case of the use of one-component systems. The directional change in composition of the polymer blend is an effective regulation method of the porous structure, morphology, crystallinity, and other characteristics of the polymeric material [2].

Presently, various blends of individual polymers are widely studied because new membrane materials possessing different functions can be made of them. For example, the use of binary mixes of polyvinyl butyral (PVB) with other polymers is of interest. It allows one to receive materials with new operational properties [3]. PVB in combination with polyvinylidene fluoride for creation of new ultrafiltration and microfiltration membranes was studied in [4]. It is known that individual PVB, which is thermoplastic, is used for the manufacturing of new and inexpensive porous membranes having thermal and chemical resistance. Polyvinylidene fluoride is a thermoplastic fluorinated polymer and it is characterized by high chemical stability in wide temperature ranges.

The purpose of this work is the IR spectroscopic investigation of the local molecular dynamics of the binary blends of polymers PVB and polyethylene glycol (PEG). We study the polymer blends with different percentage ratio of its polymers. We use the conformational probes to determine the temperature of secondary relaxation transitions of the polymer blends.

IOP Conf. Series: Journal of Physics: Conf. Series **859** (2017) 012009 doi:10.1088/1742-6596/859/1/012009

In liquid and solutions, the probe molecules are in two conformations (*trans* and *gauche*), between which a dynamic equilibrium may be established. Two analytical bands are identified in the IR spectrum of the probe, relating to the *trans* and *gauche* conformations. For their integral optical density D, the Beer–Lambert–Bouguer law is implemented:

$$D_{trans} = \alpha_{trans} l c_{trans}, \qquad D_{gauche} = \alpha_{gauche} l c_{gauche},$$

where α_{trans} and α_{gauche} are the integrated absorption coefficients of the respective bands, l is the thickness of the absorbing layer, and c_{trans} and c_{gauche} are the concentrations of *trans* and *gauche* conformations. The concentration ratio is determined by equilibrium constant K, which depends on the temperature:

$$K = c_{trans}/c_{gauche} = \exp(-\Delta G_0/RT),$$

where ΔG_0 is the difference in free energy of conformations, which can be represented as the difference between the enthalpy and entropy terms: ΔH_0 and ΔS_0 is the difference of the enthalpy and entropy of conformations, respectively. For the ratio of the optical densities of analytical absorption bands of *trans* and *gauche* conformations, we have

$$\ln(D_{trans}/D_{gauche}) = \ln(\alpha_{trans}/\alpha_{gauche}) - \Delta H_0/RT + \Delta S_0/R + \text{const}.$$

One can determine temperature T_f, at which conformational transitions of the probe in the polymer are frozen. In this case, the considered dependence in the temperature range of $T < T_g$ will consist of two branches intersecting at the temperature of T_f.

2. Experimental

PVB (Butvar B-98) and PEG (molecular weight 10,000) were manufactured by Sigma-Aldrich. PEG is an amorphous-crystalline polymer. It means a presence of both amorphous and crystalline parts. Structural formula of monomer units of these polymers are shown in figure 1.

The mixes of PVB/PEG with polymer compositions 80:20 and 60:40 were obtained by the method of physical mixing. They were prepared by dissolving the PVB and PEG in chloroform. The obtained solution was spread on KBr plates. The films were dried at room temperature for several days to evaporate the solvent completely. The absence of solvent in the samples was fixed by IR spectra.

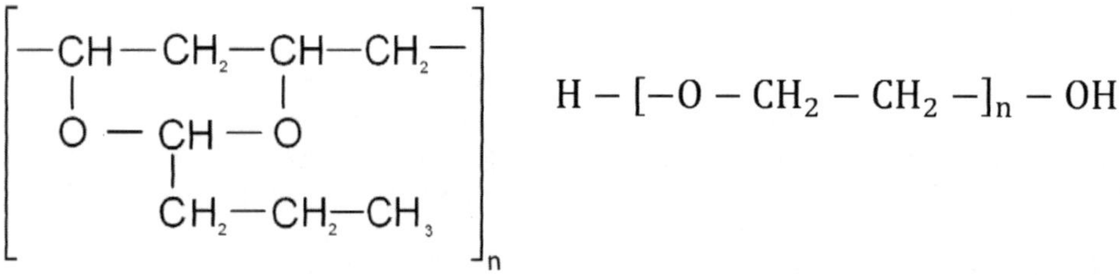

a) b)

Figure 1. Structural formula of monomer units of polyvinyl butyral (a) and polyethylene glycol (b).

Following conformationally-inhomogeneous probes were used: 1,2-diphenylethane (DPE); 1,2-dichloroethane (DCE); 1,1,2,2-tetrabromoethane (TBE), 1,1,2,2-tetrachloroethane (TCE), 1,2-diparabromophenylethane (DPBPE). The probes were chosen so that the absorption bands of the probes fall in the transparent area of the components of the binary polymer blend. However, if there was an overlapping of analytical absorption bands of probe and polymers then the subtraction of the spectrum of a blend from spectrum of system probe + polymer mixture was carried out.

Analytical absorption bands were the following conformation-sensitive absorption bands of probes at 616, 638 and 585 cm^{-1} for TBE; at 655 and 710 cm^{-1} for DCE, at 580 and 700 cm^{-1} for DPE; at 545 and 756 cm^{-1} for TCE; 538 and 518 cm^{-1} for DPBPE. TCE, DCE and TBE were introduced into polymer mixture from gaseous phase. DFE and DPBPE were introduced into the blend from the solution in chloroform. The probe concentrations were 1-3 wt. %. The concentration of the probe and the thickness of the absorbing layer were calculated by IR spectra. We used the cryostat Specac cooled by liquid nitrogen. Fourier-transform infrared spectra were recorded with a Perkin Elmer FTIR spectrometer Frontier.

3. Results and discussion

FTIR spectra of the conformational probes DCE, TBE, TCE, DPE and DPBPE in the polymer blends of PVB and PEG with polymer compositions 80:20 and 60:40 at different temperatures from 300 to 100 K were obtained. Figure 2 shows FTIR spectra of the system PVB/PEG (80:20) and DCE at three different temperatures. These spectra demonstrate a change of intensities of the conformation-sensitive (*gauche* and *trans*) absorption bands of the probe in the polymer when temperature decreases. The dependencies of logarithm of the ratio of the intensities (optical densities) of the conformationally-sensitive absorption bands upon the inverse temperatures in polymer blends showed the presence of freezing of conformational equilibria of probe's molecules in the studied temperature range. These dependencies for all probes in the polymer blends consist of two branches which cross at the freezing temperature T_f of conformational equilibrium of the probe in the polymer blend (figure 3). Besides, FTIR spectra of the probes DCE, TBE and DPE were obtained for pure PEG in the same temperature range (figure 4 and 5). Freezing of conformational mobility of the probe molecule is related to the average size of the cavity in the polymer, which is close to the size of the rotational fragment of the probe molecule $V_p^{\neq}$.

The freezing temperature T_f for the polymer blends were obtained as well as T_f for individual PEG. For individual PVB the temperatures T_f were obtained in [5] and compared with the relaxation transitions. It is shown that the temperatures T_f characterize the relaxation transitions associated with the internal rotation of C_2H_5 and C_3H_7 side groups of the polymer chain.

As to assignment of the freezing temperatures T_f to the types of mobility of the polymer chain for PEG, we consider the macromolecule of polyvinylchloride (PVC). PEG macromolecule has the structure similar to the structure of macromolecule of PVC which is also an amorphous-crystalline polymer. Besides, the monomer unit of PVC does not contain side groups. It is shown in [6] that the conformational mobility of the different probes in PVC is stopped at different temperatures, indicating the presence of several secondary relaxation transitions in PVC. When considering the local dynamics of PVC macromolecules, it is possible to distinguish three types of relaxations in the studied range of temperatures: one type is connected with torsional vibrations of CHCl fragment and two types are connected with motions of one and two monomeric units by the "crankshaft mechanism" [6]. It is assumed that there is also the local mobility of "crankshaft" in PEG. Such mechanism is also likely because the axis of rotation in PEG macromolecule is the C-O bond which has not the substituents inhibiting rotation.

Generally, the mechanism considered above for local mobility is only possible in an amorphous polymer or in amorphous regions of partially crystalline polymers. That is due to the fact that the motion of "crankshaft" type only occurs with respect to two collinear bonds. This condition is not satisfied in the crystalline regions where the sequences of methylene groups form mainly the planar *trans* conformation [7].

The freezing temperatures for the polymer blends were slightly higher than for individual PVB, namely, for DCE T_f = 210 and 195 K in PVB /PEG (60:40) and (80:20), respectively, as compared to 180 K in PVB; for TCE T_f = 230 and 235 K in these blends as compared to PVB at 225 K. For TBE the temperatures T_f are close to the mean magnitude of 240 K in both blends as compared to PVB at 235 K. For DPE T_f = 260 K in the blend (60:40) as compared to 240 K in PVB and for DPBPE T_f = 250 K in both blends as compared to 240 K in PVB.

As to the assignment of the temperatures T_f associated with relaxation transitions in the polymer blends, it is necessary to estimate the van der Waals' volume of mobile fragments (relaxators) of polymer chain of PVB and PEG by the additive scheme [8]. The evaluation gives the following volumes of the mobile fragments (relaxators) and the effective sizes of free volume elements. The relaxators C_2H_5 and C_3H_7 in PVB have volumes 27 and 44 Å^3, respectively. The volumes of fragments of polymeric chain of two "crankshaft" types, containing CH_2-CH_2-O and CH_2-CH_2-O-CH_2 groups, are equal to 44 and 64 Å^3 for PEG, respectively. Furthermore, it can be assumed that at the temperature of 190 K the mobility of H-O-CH_2 group with the volume equaled to 26 Å^3.

Thus, the IR spectroscopic method of conformational probes allows us to study the local molecular mobility in the binary polymer blends which may be used as the membranes for microfiltration and ultrafiltration.

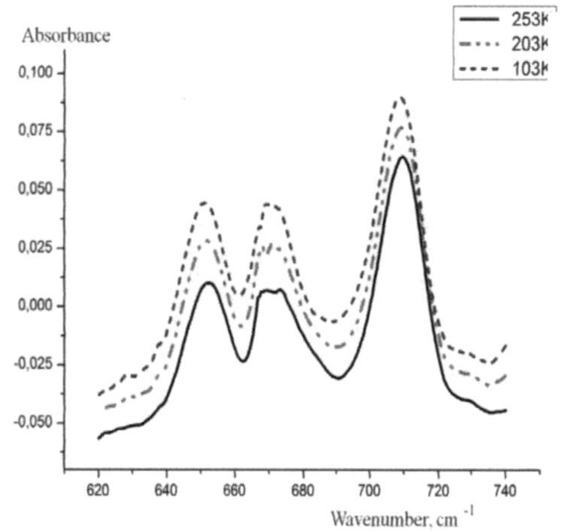

Figure 2. Fragments of FTIR spectra of DCE in PVB/PEG (80:20) at three different temperatures.

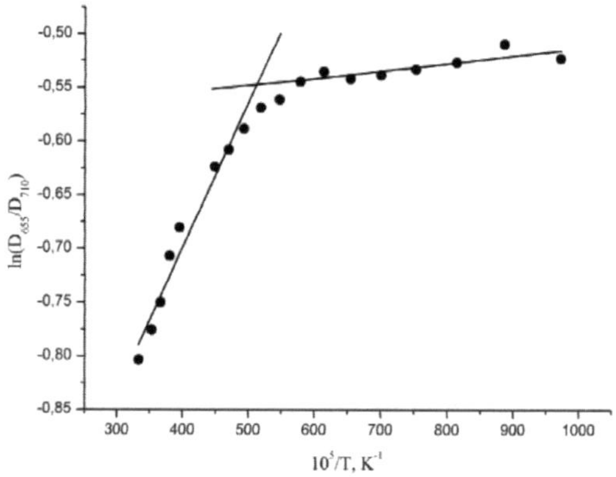

Figure 3. Dependence $\ln(D_g/D_t)$ on 1/T for DCE in PVB/PEG (80:20).

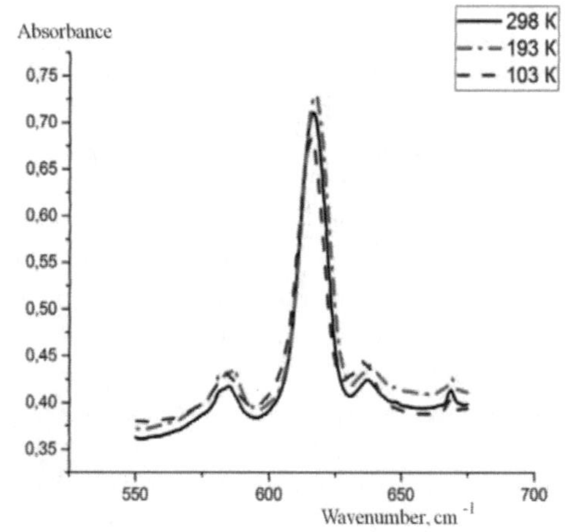

Figure 4. Fragments of FTIR spectra of TBE in PEG at three different temperatures.

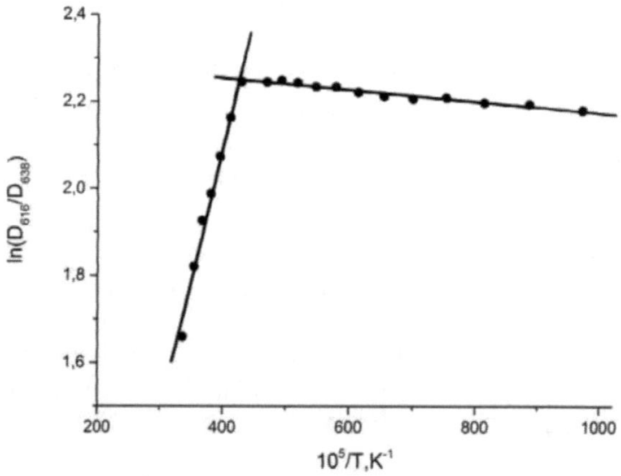

Figure 5. Dependence $\ln(D_g/D_t)$ on 1/T for TBE in PEG.

References

[1] Kamalova D I *et al* 2008 *Konformatsionnie zondi v izuchenii lokalnoi podvizhnosti polimerov* (Moscow: Fizmatkniga)
[2] Paul D and Newman C 1981 *The polymer blends* (Moscow: Mir)
[3] Ping Yun Zhang, Yan Li Wang, Zhen Liang Xu *et al* 2011 *Desalination* **278** 186-193
[4] Li Yan and Jun Wang 2011 *Desalination* **281** 455-461
[5] Kamalova D I, Shaimukhametova E R and Salahov M Kh 2013 *Uchenyye zapiski Kazanskogo universiteta* **155** 78-84
[6] Kamalova D I, Kolyadko I M and Remizov A B 2009 *Rus. J. Phys. Chem.* **83** 2278-2282
[7] Dekhant I 1972 *Infrakrasnaya spektroskopiya polimerov* (Moscow: Khimiya)
[8] Askadskiy A A and Matveyev Yu S 1983 *Khimicheskoye stroyeniye i fizicheskiye svoystva polimerov* (Moscow: Khimiya)

Photon echo in the ensemble of semiconductor quantum dots spread on a glass substrate

K R Karimullin[1,2], M V Knyazev[1], A I Arzhanov[2], L A Nurtdinova[3] and A V Naumov[1,2]

[1] Institute for Spectroscopy RAS, 108840 Moscow Troitsk 5 Fizicheskaya st.
[2] Moscow State Pedagogical University, 119991 Moscow 29 Malaya Pirogovskaya st.
[3] Kazan Federal University, Institute of Physics, 420097 Kazan 18 Kremlevskaya st.

E-mail: kamil@isan.troitsk.ru

Abstract. Simple procedure to prepare samples containing semiconductor quantum dots was developed. Test photon echo measurements in the ensemble of quantum dots spread on a glass substrate were performed to study optical dephasing processes.

1. Introduction

Photon echo is widely used in spectroscopic research of solid-state structures, liquids and gases [1]. A large number of works devoted to the using of photon echo technique to study disordered structures: polymers, glasses and molecular crystals (see, e.g., [2]). Photon echo signals generated by electronic transitions of so-called chromophores (e.g., organic molecules dyes), specially introduced into the studied matrix in a very low concentration. The optical spectra of impurity centers are extremely sensitive to the parameters of local environment, which makes it possible to use them as special probes to study the internal dynamics in various matrixes [3, 4].

Very promising in these studies could be semiconductor nanocrystals (quantum dots) with unique and controlled photophysical properties [5]. Due to the high photostability, semiconductor nanocrystals exceed the level of the quantum yield of conventional phosphors and make it easy to vary the spectrum of luminescence, which ensures their high practical value in areas such as organic photovoltaics, solar energy and quantum informatics. However, in spite of the obvious advantages of using a semiconductor nanocrystals as fluorescent probes, there are problems both with the preparation of the samples doped with quantum dots, and the detection of photon echo signal in such samples. Nowadays, a few papers on the detection of photon echo signals in the solid-state materials doped by quantum dots were published (see, e.g. [6, 7]). It should be noted, that there is a considerable interest not only to study regularities, describing the interactions of quantum dots with the environment, but also to study the photophysical properties of the quantum dots themselves. This paper is devoted to test experiments on photon echo and four-wave mixing in an ensemble of semiconductor quantum dots, spread on a glass substrate, using a developed technique in order to study the possibility of performing measurements over the wide range of low temperatures.

2. Preparation of samples

As the object of study CdSe/CdS/ZnS quantum dots (QD-light, Russia), dissolved in high concentration in toluene, were selected. Spin coating technique wasn`t used because of the large loss of the QD-solution during the creation of sample with sufficient optical density. An arrangement and procedure to spread quantum dots on a glass substrate have been developed to prepare the samples.

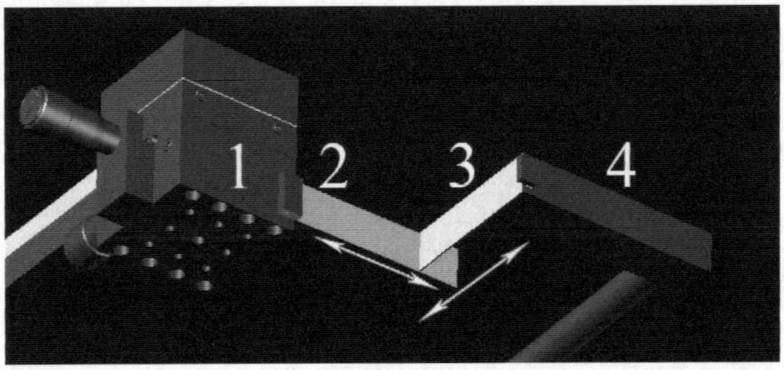

Figure 1. Schematic drawing of the setup for spreading of quantum dots on a substrate: two-axis micrometric motions (1), a glass plate (substrate) for the sample application (2), a glass plate spreading the quantum dots (3) with its support (4).

Setup (figure 1) consists of two glass plates, perpendicular to each other. One of the plates is rigidly fixed, the other is fixed on a micrometric two-axis translation stage. A small amount of a concentrated solution of quantum dots was planted on the latter glass substrate using a dispenser. The substrate with a drop is then moved towards the stationary fixed plate until it touches the drop and the solutions spreads over the edge and is being held due to surface tension. Next, translating the stage, the solution is spread layer by layer on the substrate (figure 2). Optimizing the speed of the plates motion and the distance between the plates, good quality films were produced with a high optical density. The process was visualized using a CCD- camera.

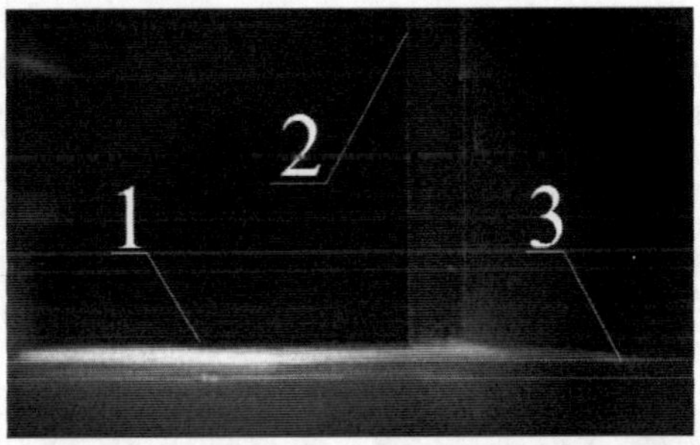

Figure 2. The image of the luminescent sample - a thin layer of quantum dots CdSe / CdS / ZnS (1) spreaded by the glass plate (2) on a glass substrate (3). The thickness of the plate (2) is 1 mm. Plate boundary circled with thin white lines to improve the contrast of the image.

3. Experimental results

The experimental data were obtained in the Molecular Spectroscopy Department of the Institute for Spectroscopy RAS by means of a unique incoherent photon echo spectrometer, developed by prof. Yu.G. Wainer in 1990 [8] and modified in recent years (see, e.g. [9, 10]).

At the heart of the setup is cavityless superluminescent light source on a dye (an ethanol solution of Rhodamine-6G) transversely pumped by second harmonic of a pulsed solid-state YAG:Nd laser (LS-2131M -10-FF, Lotis TII, Belarus). Optical scheme of the spectrometer consists of several time delay lines. The first and second form, respectively, first and second laser pulses. The delay between pulses is adjusted by changing the optical length of the first line; a second delay line is of a fixed length. Precision mechanism provided with a stepper motor can adjust the delay between the first and second pulses in the range from 5.6 fs to 4 ns with 5.6 fs minimal step. The sample is placed in the measuring chamber of an optical helium cryostat (RTI, Chernogolovka) with a high-precision control of temperature using the temperature controller Lakeshore 93C. Accurate overlapping of the focused laser beams on the test sample was performed using a confocal fluorescent visualizer [11, 12], which allows to increase the intensity of the photon echo signal generation by an order of magnitude. Signal detection was carried out with high-speed CCD camera (Cooke Corporation SensiCam HighSpeed CCD).

Figure 3 shows a decay curve of the four-wave mixing signal measured at the room temperature in an ensemble of CdSe / CdS / ZnS quantum dots spread on a glass substrate and placed inside the optical cryostat.

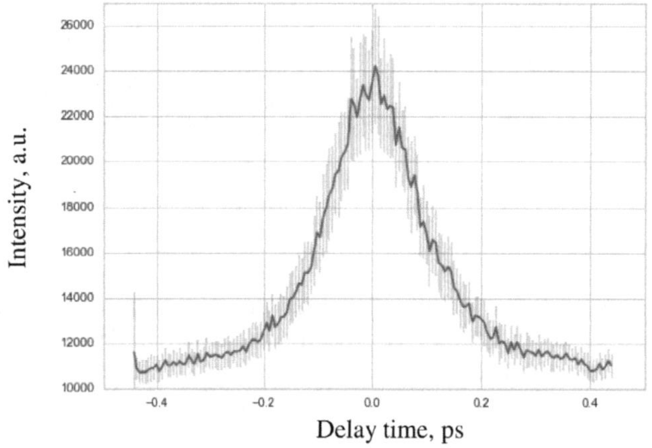

Figure 3. Four-wave mixing signal decay curve measured in an ensemble of CdSe / CdS / ZnS quantum dots on a glass substrate at the room temperature.

The decay curve characterized by a good "signal-to-noise" ratio and has a symmetrical form, inherent to a measurements at high (non-cryogenic) temperatures. At the room temperature, a decay time is comparable to the resolution time of the incoherent photon echo spectrometer (tens of femtoseconds). With decreasing temperature the shape of curve, as a rule, changes: in addition to zero delay maximum, corresponding to the four-wave mixing signal, the decline of the photon echo signal is observed in the region of positive delays. Analyzing that signal within the frameworks of various models the information about the times of relaxation processes in the investigated material is obtained.

4. Conclusion

A technique for preparing samples with semiconductor quantum dots with a sufficient optical density to perform photon echo and for-wave mixing measurements was developed. The measurements demonstrated the possibility to use these samples to study phase relaxation parameters in an ensemble of semiconductor quantum dots at low temperatures.

This work was supported by the Russian Foundation for Basic Research (projects Nos 15-32-21100-mol_a_ved – study of new optical materials for nanooptics and 14-29-07270-ofi_m – study of optical dephasing processes in quantum dots). The authors are members of leading scientific school of Russian Federation “Spectroscopy of atoms, molecules and condensed matter” (No 7035.2016.2).

References

[1] Manykin E A and Samartsev V V 1984 *Optical echo spectroscopy* (Moscow: Nauka)
[2] Vainer Yu G, Kol'chenko M A, Naumov A V, Personov R I and Zilker S J 2000 Photon echoes in doped organic amorphous systems over a wide (0.35-50 K) temperature range *J. Luminescence* **86** 265-72

[3] Naumov A V 2013 Low-temperature spectroscopy of organic molecules in solid matrices: from the Shpol'skii effect to laser luminescent spectromicroscopy for all effectively emitting single molecules *Physics-Uspekhi* **56** 605-22

[4] Magarian K A, Fedyanin V V, Karimullin K R, Vasilieva I A and Klimusheva G V 2013 Luminescence properties of CdSe nanocrystallites in cadmium alkanoate glasses *J. Phys.: Conf. Series* **478** 012007

[5] Karimullin K R and Naumov A V 2014 Dyes characterization for multi-color nanodiagnostics by phonon-less optical reconstruction single-molecule spectromicroscopy *J. Luminescence* **152** 15-22

[6] Salvador M R, Graham M W and Scholes G D 2006 Exciton-phonon coupling and disorder in the excited states of CdSe colloidal quantum dots *J. Chem. Phys.* **125** 184709

[7] Poltavtsev S V, Salewski M, Kapitonov Yu V, Yugova I A, Akimov I A, Schneider C, Kamp M, Hofling S, Yakovlev D R, Kavokin A V and Bayer M 2016 Photon echo transients from an inhomogeneous ensemble of semiconductor quantum dots *Phys. Rev.* B **93** 121304(R)

[8] Vainer Yu G and Gruzdev N V 1994 Dynamics of organic amorphous media at low-temperatures - an incoherent photon-echo study of resorufin in ethanol-d and ethanol-d(6) at 1.7-35 K 1. Experimental procedure and main results *Optics and Spectroscopy* **76** 252-8

[9] Knyazev M V, Karimullin K R and Naumov A V 2014 Using 2D-detectors in experiments on the photon echo *Proc. Samara Sci. Center RAS* **16** 13-6

[10] Karimullin K R, Knyazev M V and Naumov A V 2014 Photon echoes in an impurity polymer: New data on the low-temperature processes of phase relaxation and their relationship to the broadening of zero-phonon lines of single molecules *Bull. Russ. Acad. Sci. Phys.* **78** 1254-9

[11] Karimullin K R, Knyazev M V, Eremchev I Y, Vainer Yu G and Naumov A V 2013 A tool for alignment of multiple laser beams in pump-probe experiments *Meas. Sci. Technol.* **24** 027002

[12] Karimullin K R, Knyazev M V, Vainer Yu G and Naumov A V 2013 A luminescence visualizer for exact convergence of laser beams in photon-echo spectroscopy, four-wave mixing, and related techniques *Optics and Spectroscopy* **114** 859-62

IOP Conf. Series: Journal of Physics: Conf. Series **859** (2017) 012011 doi:10.1088/1742-6596/859/1/012011

PPLN-based photon-pair source compatible with solid state quantum memories and telecom optical fibers

I Z Latypov, A V Shkalikov, D O Akat'ev and A. A. Kalachev

E. K. Zavoisky Physical-Technical Institute, Russian Academy of Sciences, Sibirsky tract, 10/7, 420029 Kazan, Russia

bibidey@mail.ru

Abstract. We report on the realization of a tunable source of correlated photon pairs compatible with telecommunication networks and quantum memories involving dielectric crystals doped by Nd^{3+} ions. The source is based on spontaneous parametric down-conversion in a 25 mm periodically poled lithium niobate crystal pumped by 532 nm cw laser. Spectral and correlation characteristics of the corresponding heralded single-photon source compatible with quantum memories are presented.

1. Introduction

Sources of correlated photon pairs is a commonly used tool in experimental quantum optics and necessary ingredient for important quantum information protocols like quantum key distribution [1] and long-distance quantum communication via quantum repeaters [2]. In particular, for applications in long-distance quantum communication networks, it is required that quantum memories are connected to optical fibers. Therefore, nondegenerate photon-pair sources producing one photon in each pair corresponding to a telecommunication band, while another to a quantum memory absorption line, are demanded [3]. To date the most common method of creating such photon pairs is spontaneous parametric down-conversion (SPDC) [4]. Using periodically-poled crystals or waveguides makes it possible to tailor the phase matching [5] thereby allowing the creation of photons at very different wavelengths. In this work, we report on a widely nondegenerate photon-pair source compatible with Nd^{3+} doped quantum memories and with telecom optical fibers. The photon pairs are obtained by SPDC and have a wavelength near 867 nm (the resonant wavelength of Nd^{3+} ions in $YLiF_4$ crystals) and 1377 nm (the telecom E-band). The crystal $YLiF_4$:Nd^{3+} is a promising material for solid-state implementation of quantum memory protocols based on off-resonant Raman interaction [6, 7]. Recently, it was used for implementing the atomic frequency comb protocol of quantum storage [8-10]. We study the present source as a heralded single-photon source compatible with this material.

2. Basic results

In SPDC, a pump laser illuminates a material with a $\chi^{(2)}$ optical nonlinearity, creating two photons (usually called signal and idler) under the constraints of momentum and energy conservation: $k_p=k_s+k_i$ and $\omega_p = \omega_s + \omega_i$, where indexes s, i, p stand for the signal, idler and pump photons, respectively. In periodically poled crystals, the effective nonlinearity of the medium is periodically inverted by the application of an electric field with alternating directions during the crystal fabrication process. In contrast to birefringent phase-matching in bulk crystals, the quasi phase-matching conditions now involve an additional term that depends on the crystal-poling period Λ:

$$\vec{k}_p\left(\lambda_p, n_p\left(\lambda_p, T\right)\right) = \vec{k}_s\left(\lambda_s, n_s\left(\lambda_s, T\right)\right) + \vec{k}_i\left(\lambda_i, n_i\left(\lambda_i, T\right)\right) + \frac{2\pi}{\Lambda(T)}, \tag{1}$$

where $\vec{k}_{p,s,i}$ and $n_{p,s,i}$ are the wavevectors and the wavelength- and temperature-dependent refractive indices of the crystal for the corresponding wave fields. Assuming a fixed pump wavelength, equation (1) shows that the wavevectors of the signal and idler photons are temperature dependent, in particular, for a fixed emission angle. Quasi-phase-matching in periodically poled crystals allows for almost arbitrary phase-matching angles and wavelengths in comparison to the nontrivial conditions in bulk nonlinear crystals, where only birefringent and angle phase-matching is possible. In particular, periodically poled crystals can be tailored such that $\vec{k}_{p,s,i}$ are parallel to one of the crystallographic axes *X,Y* or *Z*, which allows one to avoid birefringence and utilize maximum nonlinear coefficients.

The scheme of our experimental setup is presented in figure 1. A cw neodymium laser (ALPHALAS MONOPOWER, 532 nm) pumps a 25 mm MgO:PPLN bulk crystal, provided by Labfer Ltd., with a 5% MgO-doping concentration and a poling period of 7.47μm, which gives the phasematching condition necessary to produce photon pairs at 867 nm and 1377 nm via Type 0 (*e*, *e*, *e*) SPDC process at the temperature of 85 C. This crystal has a clear aperture of 1mm × 2 mm and is anti-reflection coated on both sides for the pump, signal and idler wavelengths (see figure 1). The pump laser beam is focused with a lens of focal length 200 mm, which yields a 70 μm waist within the crystal. The generated signal (867 nm) and idler (1377 nm) photons are separated by a dichroic mirror DM1. The pump radiation is suppressed by the dichroic mirror DM2 in the idler channel and by the diffraction grating DG in the signal channel. The photons are then coupled into fibers using an aspheric lens (for the idler channel, f=11 mm) and micro objectives (for the signal channel, f=14 mm). The signal photons at 867 nm are detected by a free running Si-APD (SPCM-AQRH) with the efficiency of η_{850} =40%, dark count rate of 2 kHz and dead time of 150 ns. The idler photons are detected by an InGaAs APD (ID201, ID Quantique) in free-running mode with the efficiency of η_{1550} =10% and dead time of 12 μs. Under such conditions, the dark count rate proves to be of 2.1 kHz. To reduce broadband fluorescence background and for scanning wavelength in the signal channel, the reflecting diffraction grating mounted on a rotation servo stage is used. The spectral width of the fiber coupler input aperture is equal to 2.5±0.15 nm.

The second-order autocorrelation function of the signal field (867 nm) is measured using Braun-Twiss interferometer scheme. Signal-photon channel splits into two ones and coincidence rate between the detectors 1, 2 (R_{12}), 1, 3 (R_{13}), and 2, 3 (R_{23}) is measured simultaneously (figure 1). The coincidence rates R_{13}, R_{23} describe photon-pair generation rate. The coincidence rate R_{12} at zero time-delay between the photons propagating to detectors 1 and 2 describes the value of the second-order zero-time autocorrelation function:

$$g_2(0) = \frac{P_{12}}{P_1 * P_2}, P_{1,2} = \frac{N_{1,2}}{N_t}, \; P_{12} = \frac{N_{12}}{N_t}, \tag{2}$$

where P_{12} is the probability of joint detection on the detectors 1 and 2; P_1, P_2 is the probability of detection on the detectors 1 and 2, respectively; $N_{12} = R_{12}$; N_1=R_1; N_2=R_2; N_t is the total normalized number of pulses coming to the channels 1 and 2. Normalization of N_t is carried out on total losses in the optical tract (optical elements, fiber coupling) and quantum efficiency of detectors. Figure 2 illustrates a linear dependence of $g_2(0)$ on the pump power. After optimization, we obtained the following parameters of the photon-pair production: the photon-pair generation rate, considering the general losses in the experimental setup, is equal to 160 kHz*mW/sec, and $g_2(0)$=0.004±0.001 for the pump power of 2 mW. Spectral width of the signal photons, which was measured using the diffraction grating DG, proves to be of 0.15±0.3 nm. The width of the correlation function measured via time-to-digital converter ID800 is of 330±81 ps. Figure 3 illustrates the possibility of the temperature tuning of the single-photon wavelength. To simulate the experimental results, we used modified Sellmeier equations for the temperature-dependent refractive indices $n(\lambda, T)$ from [11].

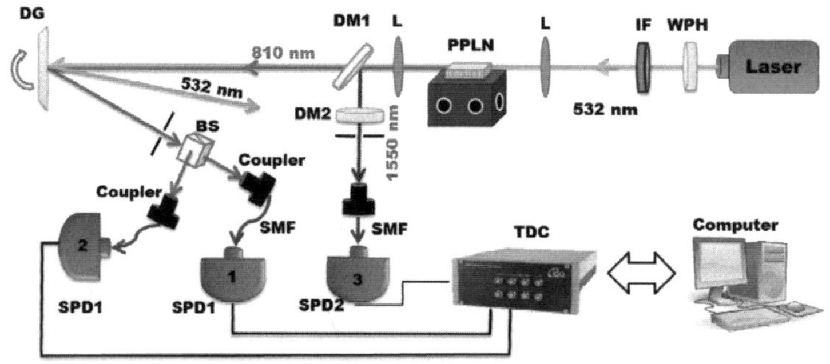

Figure 1. Schematic of the experimental setup. Laser is the cw neodymium laser ALPHALAS MONOPOWER-532-100-SM; L is a coated lens; DM1 is a dichroic mirror with 98% transmission at 850 nm and 94% reflection at 1550 nm; IF is an interference filter for the pump field at 532 nm; WHP is a half-wave plate at 532 nm; DM2 is a dichroic mirror with 95% transmission at 1550 nm and 91% reflection at 532 nm; PPLN is the periodically poled lithium niobate placed in a thermostat for crystal temperature stabilization; DG is a reflective diffraction grating with the angular dispersion of 3.2 nm/mrad placed on a rotation stage; BS is a nonpolarizing 50/50 beam splitter; Coupler is a cage fiber coupler system; SMF is a single-mode fiber; SPD1, SPD2 are Perkin Elmer single-photon detectors; SPD3 is the single photon detector ID210; ID800 is the eight-channel time-to-digital converter with the time resolution of 81 ps.

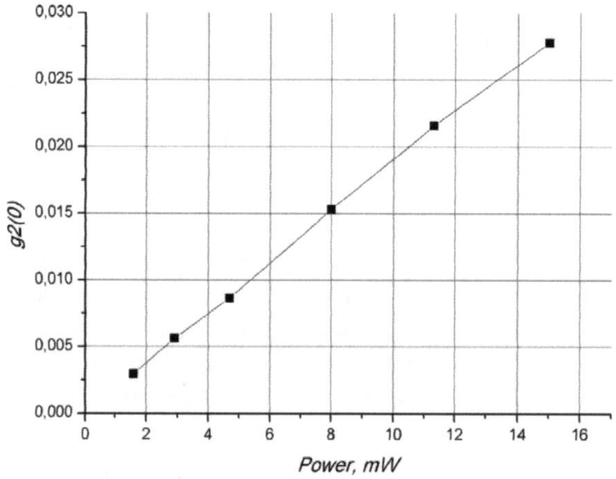

Figure 2. The dependence of the second-order zero-time autocorrelation function $g^{(2)}(0)$ on the pump power.

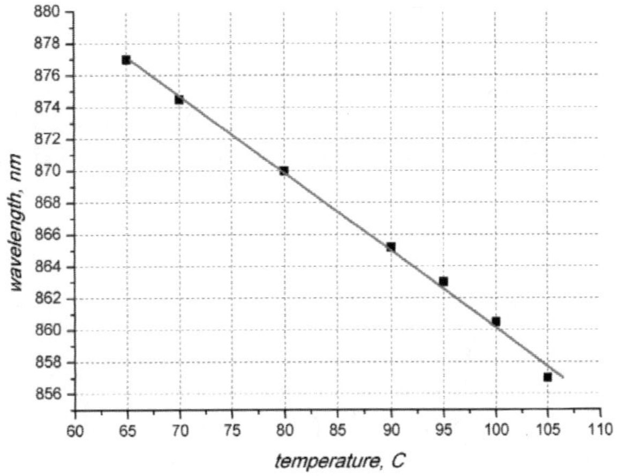

Figure 3. Wavelength of the signal photons vs temperature of the PPLN crystal. Solid line is the result of theoretical calculation using equation (1).

3. Conclusion

We have studied spontaneous parametric down-conversion in a periodically poled lithium niobate crystal wherein correlated photon pairs compatible with telecommunication networks (wavelength 1377 nm) and Nd^{3+} doped quantum memories (wavelength 867 nm) are produced. The basic features of the corresponding heralded single-photon source compatible with such quantum memories are studied. In particular, the second-order zero-time autocorrelation function for the field generated at 867 nm is measured as a function of the pump power, and possibilities of wavelength tuning are demonstrated. The present source may be used as a basis for developing a narrowband variant based on the cavity-enhanced spontaneous parametric down-conversion, which can produce photons interacting effectively with the $YLiF_4$:Nd^{3+} crystal.

Acknowledgments

This research was supported by the Russian Science Foundation (Grant No. 14-12-00806).

References

[1] Ekert A K 1991 *Phys. Rev. Lett.* **67** 661-663
[2] Briegel H-J, Dür W, Cirac J I and Zoller P 1998 *Phys. Rev. Lett.* **81** 5932–5935
[3] Sangouard N, Simon C, de Riedmatten H and Gisin N 2011 *Rev. Mod. Phys.* **83** 33–80
[4] Mandel L and Wolf E 1995 *Optical Coherence and Quantum Optics* (Cambridge University Press, New York) p 1166
[5] Tanzilli S, Martin A, Kaiser F, DeMicheli M P, Alibart O and Ostrowsky D B 2012 *Laser & Photon. Rev.* **6** 115–143
[6] Moiseev S A and Tittel W 2011 *New Journal of Physics* **13** 063035
[7] Zhang X, Kalachev A, Kocharovskaya O 2013 *Phys. Rev.* A **87** 013811(1-9)
[8] Zhou Z-Q, Wang J, Li C-F and Guo G-C 2013 *Scientific Reports* **3** 2754(1-4)
[9] Akhmedzhanov R A, Gushchin L A, Kalachev A A, Korableva S L, Sobgayda D A and Zelensky I V 2016 *Laser Physics Letters* **13** 015202(1-5)
[10] Akhmedzhanov R A, Gushchin L A, Kalachev A A, Nizov N A, Nizov V A, Sobgayda D A and Zelensky I V 2016 *Laser Physics Letters* **13** 115203(1-5)
[11] Shen H Y, Xu H, Zeng Z D, Lin W X, Wu R F and Xu G F 1992 *Applied Optics* **31** 6695-6697

A polyphase sequences with low autocorrelations

A N Leukhin[1,2]

[1]Physics and Mathematics Department, Mari State University, 1, sq. Lenin, Yoshkar-Ola, Mari El Rep, 424000, Russia
[2]Computer science Department, Volga State University of Technology, 3, sq. Lenin, Yoshkar-Ola, Mari El Rep, 424000, Russia

E-mail: leukhinan@list.ru

Abstract. This paper gives numerical method for unimodular sequences design with low peak sidelobe (PSL) level of aperiodic autocorrelation function (AACF). The result of numerical optimization for design of unimodular sequences of length $N = 1024$ with smallest today $PSL \approx 2.618$ is shown. Such sequences are useful for resolution and parameter estimation in radar applications due to their large signal-to-noise ratio.

1. Introduction

One of a problem of sequence design is to determine a long unimodular sequence for which the absolute values of the aperiodic autocorrelations are collectively small as possible.

Let us determine an unimodular sequence in the manner of:

$$\vec{U} = (u_0, u_1, .., u_{N-1},), \tag{1}$$

where a code element $u_n = \exp(\sqrt{-1} \cdot \varphi_n)$, a phase φ_n is any value from $\varphi_n = [0, 2\pi]$, modulo of all code elements $|u_n| = 1$ are strictly equal to 1.

If $\varphi_n = \{0, \pi\}$ phase takes only two values then unimodular sequence is a binary sequence with alphabet $u_n = \{-1,1\}$. If phase $\varphi_n = \frac{2\pi}{M} \cdot k$, $k = 0,1,...,M-1$ then a unimodular sequence is M-ary sequence ($M = 3$ - ternary sequence, $M = 4$ quaternary sequence and so on). In general case when $\varphi_n = [0, 2\pi]$ unimodular sequences (1) is also called as polyphase sequence.

Aperiodic ACF $\vec{C} = (c_0, c_1, ..., c_{N-1})$ of the unimodular sequence can be written in the form

$$c_\tau = \sum_{n=0}^{N-1-\tau} u_{n+\tau} \cdot \overline{u_n}, \ \tau = 0,1,...,N-1, \tag{2}$$

where $\overline{u_n}$ is complex conjugated value for u_n.

There are two criteria for the optimality of binary sequences with low levels of AACF. The first is the minimax criterion. The peak sidelobe PSL

$$PSL\left(\vec{C}\right) = \max_{1 \le \tau \le N-1} \left|c_\tau\right| \tag{3}$$

must be minimal

$$MPS = \min_{S} PSL. \tag{4}$$

The second is the *MF* (merit factor) coefficient characterizing the ratio of the main frame energy to that of the sidelobes of AACF

$$MF\left(\vec{C}\right) = \frac{N^2}{2\sum_{\tau=1}^{N-1} \left|c_\tau\right|^2}. \tag{5}$$

The optimum binary sequences of a given length N has the highest value of the MF coefficient.

One of the problems of sequence design is to determine long unimodular sequences for which the absolute values of the aperiodic autocorrelations are collectively small as possible. The last sidelobe of AACF is always equal by modulo to 1. So the optimal aperiodic sequence has minimum $PSL=1$. Barker [1] firstly constructed binary sequences with the lengths $N = 2,3,4,5,7,11,13$ for which all non-trivial aperiodic autocorrelations are at most 1 in magnitude. Turin [2] showed that no Barker sequence of length greater than 13 exists for even lengths. Also in [3] it was shown that if a Barker sequence of length $N > 13$ exist, then either $N = 3979201339721749133016171583224100$, or $N > 4 \cdot 10^{23}$.

Another way to construct an unimodular sequence with $PSL=1$ is to use a non-binary alphabet. Ternary Barker codes [4] (three phases) have only been found up to $N=9$, quarternary Barker codes (four phases) have only been found up to $N=15$ [5], sextic Barker codes (six phases) have only been found up to $N=18$ [6]. Golomb and Scholtz [6] called sequences with non-binary alphabet as generalized Barker sequences if they have PSL equal or less than 1.

Friese and Zottmann [7] found generalized Barker sequences up to $N=31$. Later Friese [8] continued generalized Barker sequences up to $N=36$. Brenner [9] found generalized Barker sequences up to $N=45$. Borwein and Ferguson [10] performed search up to $N=63$. Nunn [11] found generalized Barker sequences for the length $N=64$. Nunn and Coxson [12] found generalized Barker sequences for the lengths $N=[65,70]$ and $N = 72,76,77$. This are still longest lengths of known polyphase Barker sequences. So a numerical optimization search for generalized Barker sequences are allowed to construct optimal sequences for limited lengths.

There are some analytical constructions of polyphase sequences with good aperiodic autocorrelations. The Frank sequence [13] has the lowest $PSL=\sqrt{N}/\pi$ level of AACF between all known regular constructions of unimodular sequences. For Frank sequence of length $N=1024$ the PSL level is equal to $PSL=10.202$.

In this paper the new numerical method for design of polyphase sequence with low PSL level of AACF is offered. Within the frame of developed method it were constructed the polyphase sequences with today known lowest PSL level of AACF for the lengths N=128, 256, 512, 1024, 2048, 4096. The constructed polyphase sequence of length N=1024 with $PSL \approx 2.618$ is presented.

2. Numerical method for design of polyphase sequences with low PSL level

For the generalized Barker sequences as well as for Barker binary sequences AACF is determined as

$$\left|c_\tau\right| = \begin{cases} N, & when\ \tau = 0, \\ \le 1, & when\ \tau = 1,2,\ldots,N-1, \\ 1, & when\ \tau = N-1. \end{cases} \tag{6}$$

Let us equalized all non trivial sidelobes of AACF:

$$c_0 = N\,,\ \left|c_\tau\right| = 1\,,\ \tau = 1,2,\ldots,N-1\,. \tag{7}$$

First it is possible to get the system of trigonometric equations:

$$\left[\sum_{n=0}^{N-1-\tau}\cos(\varphi_n - \varphi_{n+\tau})\right]^2 + \left[\sum_{n=0}^{N-1-\tau}\sin(\varphi_n - \varphi_{n+\tau})\right]^2 = 1\,,\ \tau = 1,2,\ldots,N-1\,. \tag{8}$$

And it is possible to reduce this system to the form:

$$2\sum_{n=0}^{N-2-\tau}\sum_{k=n+1}^{N-1-\tau}\cos(\varphi_{\tau+n} - \varphi_n - \varphi_{\tau+k} + \varphi_k) = \tau - N\,,\ \tau = 1,2,\ldots,N-1\,. \tag{9}$$

For normalized unimodular Barker sequence it is possible to find a solution of system (9) in the manner of phase vector

$$\vec{\Psi} = (\varphi_0 = 0, \varphi_1 = 0, \varphi_2, \ldots, \varphi_{N-1}), \tag{10}$$

using the next iterative method.

1. Get initial phase vector $\vec{\Psi}^{(0)}$ with good aperiodic autocorrelations from known binary or polyphase sequences.

2. In each iteration this vector will be changed to the next value

$$\vec{\Psi}^{(i+1)} = \vec{\Psi}^{(i)} + \Delta\vec{\Psi}\,, \tag{11}$$

where increment $\Delta\Psi$ of initial vector is determined using Levenberg-Marquardt algorithm [14] as

$$\Delta\vec{\Psi} = [J^T J + \lambda_k diag(J^T J)]^{-1} J^T (\vec{y} - f(\vec{\Psi}^{(i)}))\,, \tag{12}$$

where J is an Jacobian matrix of the function $f\left(\vec{\Psi}\right)$, $\vec{y}$ is a vector of phase components, λ_k is a damping parameter which is determined for each k th step.

Jacobian matrix J of size $N \times R$ is presented as

$$J = \begin{bmatrix} \dfrac{\partial f\left(\vec{\Psi}\right)_1}{\partial \vec{\Psi}_1} & \cdots & \dfrac{\partial f\left(\vec{\Psi}\right)_1}{\partial \vec{\Psi}_R} \\ \vdots & \ddots & \vdots \\ \dfrac{\partial f\left(\vec{\Psi}\right)_N}{\partial \vec{\Psi}_1} & \cdots & \dfrac{\partial f\left(\vec{\Psi}\right)_N}{\partial \vec{\Psi}_R} \end{bmatrix}, \tag{13}$$

IOP Conf. Series: Journal of Physics: Conf. Series **859** (2017) 012012 doi:10.1088/1742-6596/859/1/012012

where

$$\frac{\partial f\left(\vec{\Psi}\right)_i}{\partial \vec{\Psi}_r} = \frac{f\left(\vec{\Psi}+h\right)_i - f\left(\vec{\Psi}\right)_i}{h}, \tag{14}$$

where h is differential interval which is determined from

$$100\cdot\varepsilon < h < 1000\cdot\varepsilon, \tag{15}$$

where ε is computing tolerance.

To evaluate the result of iteration it is possible to minimize the function

$$F\left(\vec{\Psi}\right) = \sum_{i=0}^{N-1} f^2\left(\vec{\Psi}\right)_i, \tag{16}$$

where $F\left(\vec{\Psi}\right) \le \varepsilon$.

To construct unimodular with low PSL level of AACF it is necessary to solve the system

$$f\left(\vec{\Psi}\right)_m \le PSL,\ m = 1,2,\ldots,N-1 \tag{17}$$

The function $F\left(\vec{\Psi}\right)$ is determined as

$$F\left(\vec{\Psi}\right) = \sum_{i=0}^{N-1} \begin{cases} k^2 & \mathit{if}\ f\left(\vec{\Psi}\right)_i \le PSL,\ k < \varepsilon, \\ f^2\left(\vec{\Psi}\right)_i & \mathit{otherwise}. \end{cases} \tag{18}$$

Taking into account that $PSL = const$ the value of Jakobian matrix (13) is not changed.

3. Polyphase sequence of length *N*=1024 with low PSL level

Let us demonstrate the results of unimodular sequence design with low PSL level of AACF. The phase vectors of unimodular sequence of length $N = 1024$ with $PSL \approx 2.618$ is shown in Table 1.

Table 1. Polyphase sequence with low PSL level of AACF

Phase vector
0; 0; 0.502354; 1.20153; 1.92004; 2.82929; 3.69897; 4.63371; 5.39091; 0.0202442; 0.919502; 1.96029; 2.96028; 3.87082; 4.81008; 5.63301; 0.285997; 1.45774; 2.33486; 3.32064; 4.23246; 5.24315; 6.15208; 0.816183; 1.70247; 2.78738; 3.72815; 4.60078; 5.54351; 0.345064; 1.29089; 2.24139; 3.15071; 4.0197; 4.97496; 5.9496; 0.518607; 1.36291; 2.21849; 3.13785; 3.88395; 4.59976; 5.33999; 5.96704; 0.200324; 0.71184; 1.22427; 1.81442; 2.44167; 3.0837; 3.85676; 4.78726; 5.65116; 0.254958; 1.18702; 1.95481; 2.7541; 3.49985; 4.36427; 5.3646; 6.16369; 0.527665; 1.38709; 2.07912; 2.57514; 2.97313; 3.58478; 4.30928; 4.81985; 5.34005; 6.00731; 0.388543; 1.04299; 1.62576; 2.28652; 2.95024; 3.47678; 3.83596; 4.22526; 4.44534; 5.09303; 5.42166; 5.87191; 0.106682; 0.790245; 1.59574; 2.3443; 2.88718; 3.55709; 4.1525; 4.85734; 5.42407; 5.95296; 0.273906; 0.732587; 1.308; 1.80114; 2.40261; 2.97433; 3.45991; 3.79059; 4.09671; 4.15761; 4.26068; 4.42331; 4.61476; 5.07355; 5.61661; 6.06421; 0.186735; 0.692891; 1.21266; 1.52511; 1.80412; 2.07663; 2.38657; 2.69133; 3.06445; 3.37076; 3.73742; 4.06492;

IOP Conf. Series: Journal of Physics: Conf. Series **859** (2017) 012012 doi:10.1088/1742-6596/859/1/012012

Table 1. Continued

Phase vector
4.36454; 4.70337; 5.09485; 5.53437; 5.99737; 0.224282; 0.595864; 1.07076; 1.39921; 1.49087; 1.62497; 1.77112;
1.96106; 2.16928; 2.3672; 2.55729; 2.49787; 2.41681; 2.38961; 2.53531; 2.79427; 2.95409; 3.19493; 3.39461;
3.51714; 3.74236; 3.94372; 4.14388; 4.29585; 4.27096; 4.21631; 4.15811; 4.35935; 4.60465; 5.10349; 5.57762;
5.86299; 5.99446; 6.09942; 6.14928; 6.22609; 0.120384; 0.128815; 6.21819; 5.92585; 5.91551; 5.95087; 5.98603;
6.11723; 6.23797; 6.23785; 6.10519; 5.99681; 5.73763; 5.35949; 4.90509; 4.88344; 5.1286; 5.21629; 5.40747;
5.50044; 5.56862; 5.53187; 5.54062; 5.67782; 5.79024; 5.7476; 5.58928; 5.39718; 5.27009; 5.22447; 5.05819;
4.753; 4.53037; 4.40691; 4.46705; 4.33115; 4.06859; 3.79666; 3.55531; 3.35602; 3.16981; 2.82852; 2.47885;
2.258; 2.13918; 1.9809; 1.50054; 0.883567; 0.457827; 0.379224; 0.284925; 0.206963; 6.28318; 6.18734; 6.1296;
5.97543; 5.58132; 5.08034; 4.84595; 4.97211; 5.00047; 4.7204; 4.356; 4.21777; 3.8099; 3.03456; 2.25862;
1.89082; 1.60507; 1.20917; 0.672054; 0.202243; 5.93011; 5.10184; 4.55087; 4.20011; 3.97975; 3.70049; 3.42496;
3.19689; 2.7756; 2.35198; 1.70569; 1.26031; 0.919614; 0.539523; 0.0675575; 5.79401; 5.31055; 4.98951; 4.59964;
4.22913; 4.06072; 3.77018; 3.67144; 3.28832; 2.49429; 1.90632; 1.42997; 0.623824; 5.98061; 5.2342; 4.56714;
3.87782; 3.49148; 3.09849; 2.46502; 1.54987; 1.02056; 0.500525; 5.97924; 5.3797; 4.92743; 4.47883; 3.72699;
3.01527; 2.35232; 1.63641; 0.940282; 0.542341; 6.1777; 5.31659; 4.6542; 4.21414; 3.75031; 3.0864; 2.37262;
1.79142; 1.18613; 0.260809; 5.6344; 4.99509; 4.19189; 3.57233; 2.95972; 2.20488; 0.870947; 6.08931; 5.36088;
4.68375; 3.91472; 3.14791; 2.12121; 0.986391; 0.147358; 5.51336; 4.71236; 3.983; 3.34281; 2.53863; 1.75445;
1.02967; 0.378665; 6.03657; 5.33095; 4.61569; 3.85434; 2.88142; 1.982; 0.866447; 5.9708; 4.86444; 3.7146;
2.63872; 1.71888; 0.70809; 6.06068; 5.17752; 4.29313; 3.3078; 2.45619; 1.46145; 0.325924; 5.45321; 4.21432;
3.15225; 2.27512; 1.32756; 0.233402; 5.49194; 4.46695; 3.42973; 2.44201; 1.48855; 0.549428; 5.90659; 4.91751;
3.77637; 2.77551; 1.88845; 0.822135; 6.0697; 4.73676; 3.38558; 2.3446; 1.26202; 6.24854; 5.02724; 3.5013;
1.93814; 0.699791; 5.96722; 4.84237; 3.61362; 2.41359; 1.35276; 0.410833; 5.57689; 4.23597; 2.87344; 1.59944;
0.33196; 5.26751; 3.86471; 2.41162; 1.14269; 0.159388; 5.35111; 4.20711; 3.15517; 2.08357; 0.983953; 6.07742;
4.88108; 3.62731; 2.41498; 0.884876; 5.41933; 3.72805; 2.22556; 0.949445; 5.95935; 4.76135; 3.45452; 1.90793;
0.306553; 5.07704; 3.6153; 2.17925; 0.952911; 6.04621; 4.43943; 2.88599; 1.67662; 0.47217; 5.34851; 3.65971;
2.07175; 0.695655; 5.96749; 4.68454; 3.21404; 1.6942; 0.407981; 5.31746; 3.8541; 2.21712; 0.442035; 5.16768;
3.61885; 2.09628; 0.396418; 4.97881; 3.41458; 1.93426; 0.569078; 5.45451; 3.83595; 2.22293; 0.571216; 5.10099;
3.33342; 1.7045; 0.191741; 4.80273; 3.01201; 1.33969; 6.09583; 4.50597; 2.92847; 1.36228; 6.04057; 4.49714;
2.96261; 1.47915; 6.179; 4.51416; 2.67525; 0.79034; 5.40426; 3.56691; 1.67767; 6.16273; 4.37099; 2.53615;
0.628923; 4.92682; 3.04957; 1.38208; 6.03267; 4.37096; 2.69374; 0.991824; 5.05763; 3.03864; 1.21748; 5.8425;
4.06614; 2.3323; 0.650326; 5.1938; 3.32709; 1.4071; 6.00664; 4.31012; 2.6111; 0.692052; 5.27876; 3.63373;
1.8737; 6.11484; 3.64952; 1.41515; 5.73363; 3.77104; 1.78035; 6.16872; 4.27227; 2.40542; 0.540733; 4.9182;
2.86956; 0.818216; 4.88834; 2.94725; 1.26308; 5.87289; 3.58876; 1.44681; 5.55369; 3.57148; 1.4988; 5.70009;
3.57253; 1.5283; 5.78311; 4.09139; 2.14913; 0.169011; 4.54514; 2.56892; 0.360225; 4.40785; 2.23173; 0.167578;
4.31022; 2.0866; 6.00154; 3.77144; 1.46141; 5.54263; 3.32267; 1.29855; 5.44196; 3.25176; 0.789603; 4.8162;
2.71639; 0.658682; 4.7301; 2.70422; 0.518759; 4.3593; 2.05452; 6.07582; 3.90059; 1.96468; 0.00193823; 4.14469;
2.21808; 0.154306; 4.09761; 1.71006; 5.41743; 3.02856; 0.767376; 4.63414; 2.13215; 5.93975; 3.712; 1.37134;
5.18611; 2.76754; 0.387408; 4.33911; 1.87334; 5.89454; 3.92613; 1.9297; 5.4889; 2.86394; 0.358557; 4.11265;
1.64256; 5.55618; 3.31513; 0.973113; 4.80315; 2.30787; 6.1579; 4.00375; 1.7763; 5.81162; 3.41949; 0.939669;
4.57336; 2.01461; 5.70268; 3.03947; 0.441085; 4.16239; 1.62733; 5.13931; 2.47364; 6.21918; 3.73881; 1.35505;
4.95259; 2.21752; 5.78962; 3.16384; 0.657633; 4.4455; 2.03054; 5.7324; 3.14769; 0.503428; 4.03502; 1.36366;
5.15027; 2.69588; 0.0446304; 3.74166; 1.16448; 4.9458; 2.64849; 6.2297; 3.49115; 0.757837; 4.32556; 1.55526;
4.78233; 1.53037; 4.95154; 2.20426; 5.84698; 3.00614; 0.132208; 3.82307; 1.32595; 5.01059; 2.17431; 5.50732;
2.78633; 6.26505; 3.6004; 0.791799; 4.16039; 1.3977; 4.8141; 2.01758; 5.47761; 2.65599; 0.19993; 3.95213;
1.26483; 4.49792; 1.48917; 4.86481; 1.94563; 5.1736; 2.24738; 5.39943; 2.44564; 5.80891; 2.75929; 5.86235;
3.17822; 0.301423; 3.84727; 0.931741; 4.28362; 1.31654; 4.51313; 1.32514; 4.51412; 1.42112; 5.043; 2.08082;
5.05866; 1.79208; 5.16423; 2.29399; 6.176; 3.40431; 0.260793; 3.46011; 0.504444; 3.80086; 0.895334; 4.31402;
1.24963; 4.35118; 1.17335; 4.20013; 1.00479; 4.03301; 0.713638; 3.77997; 0.654921; 3.71372; 0.567244; 3.76953;
0.678644; 3.85953; 0.705114; 3.72996; 0.544395; 3.71117; 0.683334; 4.04536; 0.808474; 3.75232; 0.473518;
3.49029; 0.149825; 3.22501; 0.00328948; 3.24126; 0.342262; 3.67485; 0.533612; 3.74279; 0.449832; 3.3833;
6.09541; 2.62453; 5.5096; 2.12778; 5.02985; 1.60932; 4.48535; 1.11066; 3.95758; 0.417916; 3.20144; 6.13906;
2.87825; 6.05669; 2.93383; 5.79826; 2.21795; 5.11564; 1.89221; 5.00433; 1.77881; 4.66489; 0.999657; 3.69976;
0.381886; 3.37105; 0.142951; 3.19872; 6.04741; 2.49242; 5.1977; 1.55786; 4.31994; 0.607877; 3.38532;

Table 1. Continued

Phase vector
0.00754261; 2.63765; 5.20961; 1.73336; 4.51474; 0.923641; 3.60507; 0.197534; 2.98919; 5.52975; 1.98271;
4.95358; 1.45181; 3.93134; 0.328216; 3.05218; 5.6772; 1.73593; 4.46957; 1.06321; 3.89067; 0.499236; 3.52607;
0.11513; 2.70204; 5.40739; 1.78496; 4.38786; 0.371253; 2.39075; 4.97831; 1.25471; 3.70422; 6.17411; 2.47566;
5.12767; 1.42764; 3.9823; 0.424739; 3.01709; 5.44318; 1.51564; 3.76009; 6.05319; 2.30764; 4.7922; 0.998507;
3.86223; 0.327898; 2.85359; 5.35658; 1.59009; 4.16952; 0.494806; 3.0394; 5.83728; 2.28342; 4.7138; 0.795657;
3.03044; 5.43815; 1.58195; 3.84128; 6.06143; 2.21524; 4.66106; 0.675068; 2.75005; 4.85761; 0.829752; 3.19068;
5.61724; 1.62938; 4.09179; 0.515883; 2.94874; 5.17966; 0.975193; 3.26427; 5.67423; 1.86048; 4.22829; 0.209599;
2.4759; 4.83827; 0.892902; 3.12707; 5.23861; 1.20453; 3.55267; 6.06415; 2.087; 4.24625; 0.062007; 2.19628;
4.44342; 0.321177; 2.2364; 4.14851; 6.17563; 2.02446; 4.15638; 6.2145; 1.92505; 3.98677; 5.95249; 1.71595;
4.25258; 0.546689; 2.96764; 4.9577; 0.665928; 2.93606; 4.92156; 0.67494; 2.79515; 4.85427; 0.701538; 2.98706;
5.29646; 1.26057; 3.15081; 5.17023; 0.759423; 2.53509; 4.57017; 0.23538; 1.94258; 3.66688; 5.54366; 1.18234;
3.29025; 5.45102; 1.34563; 3.37292; 5.32795; 0.852491; 2.62745; 4.47614; 0.115401; 2.07925; 3.98251; 5.854;
1.45503; 3.38365; 5.65778; 1.60797; 3.5862; 5.44817; 1.07379; 2.94247; 4.83569; 0.470284; 2.43966; 4.36761;
6.0047; 1.51092; 3.05775; 4.56804; 6.27966; 1.70483; 3.45485; 5.30237; 0.811618; 2.57092; 4.33989; 6.08141;
1.45605; 3.25501; 4.86168; 0.364071; 2.22197; 4.01206; 5.73798; 1.16769; 2.86524; 4.35037; 6.05321; 1.4595;
3.30909; 5.39471; 1.25816; 3.21578; 4.91909; 0.3442; 1.92275; 3.57578; 4.86478; 0.0126062; 1.6544; 3.30011;
4.96548; 0.236804; 1.80352; 3.15581; 4.61615; 6.09285; 1.27854; 2.70783; 4.11869; 5.60728; 1.01136; 2.60605;
4.07671; 5.91797; 1.47639; 3.27301; 4.74741; 6.27354; 1.44482; 2.88253; 4.34709; 5.7135; 0.952381; 2.53795;
4.06874; 5.50084; 0.859424; 2.42511; 4.02902; 5.31923; 0.524747; 1.90397; 3.22119; 4.28486; 5.59865; 0.66489;
1.92816; 3.12248; 4.43353; 5.71015; 0.699963; 1.8664; 3.23422; 4.59612; 6.02073; 1.28789; 2.82486; 4.33882;
5.85735; 1.081; 2.52077; 3.94776; 5.18859; 0.235188; 1.50979; 2.8474; 3.99225; 5.14361; 6.28243; 1.24883;
2.54353; 3.75667; 4.75961; 5.80364; 0.613861; 1.86026; 3.05527; 4.21603; 5.39571; 0.257611; 1.38533; 2.50064;
3.68705; 4.91227; 6.06777; 0.841194; 1.86114; 2.97253; 4.07344; 5.30323; 0.17064; 1.3602; 2.5224; 3.8304;
5.06502; 6.26166; 1.14431; 2.37397; 4.04471; 5.27239; 0.336593

4. Conclusion

We constructed unimodular sequences of large lengths which are pretty close to generalized Barker codes. The results of unimodular sequence optimization for minimizing the PSL level of AACF are not finally finished. It is possible to decrease the PSL level and to get sufficiently smaller values which are should be closer to $PSL = 1$.

Acknowledgments

This work is supported by the grants: grant of Russian Foundation of Basic Research 12-07-00552 and projects of Russian Ministry of Education and Science 2.2226.2017/HR and 02.G25.31.0204.

References

[1] Barker R H 1953 *Comm. Theory* ed W Jackson (New-York: Academic Press)
[2] Turin R 1963 *Proceedings of the IEEE* **51** 1256
[3] Borwein P and Mossinghoff M J 2014 *LMS J. Comput. Math* **17** 24-32
[4] Jr de Long D F 1959 (MIT Lincoln Lab. Lexington Massachusetts Group Report) pp 47-48
[5] Welti G R 1960 *IRE Transaction on Information Theory* **6** 400-408
[6] Golomb S W and Scholtz R A 1965 *Trans. on Inf. Theory* **11** 533-537
[7] Friese M and Zottmann H 1964 *Electronics Letters* **30** 1930–1931
[8] Friese M 1996 *IEEE Transactions on Information Theory* **42** 1248–1250
[9] Brenner A R 1998 *Electronics Letters* **34** 1576–1577
[10] Borwein P B and Ferguson R A 2005 *IEEE Trans. Inform. Theory* **51** 1564–1567
[11] Nunn C J 2005 *Proc. of the 2005 IEEE Int. Radar Conference* (Washington D C) pp 190-194
[12] Nunn C J and Coxson G E 2009 *IEEE Trans. on Aerosp. and El. Syst.* **45** 775-781
[13] Frank R L 1963 *IEEE Trans Inform Theory* **9** 43-45
[14] Nocedal J and Wright S J 2006 *Numerical Optimization* 2nd ed (Berlin, Germany: Springer)

Quantum fluctuations in semiconductor quantum dots and their contributions to the self-energy functions of exciton states

A A Mutygullina, M A Khamadeev, D O Blum and A H Shirdelhavar

Optics and Nanophotonics Department, Institute of Physics, Kazan Federal University, 18 Kremlevskaya St, Kazan 420008, Russia

E-mail: Renat.Gainutdinov@kpfu.ru

Abstract. Influence of quantum fluctuations in a system consisting of a quantum dot and the reservoir of acoustic phonons on processes in which the quantum dot takes part is investigated. Under some conditions this influence is shown to be very strong. We find a contribution from the quantum fluctuations to the self-energy function of the exciton coupled to the quantum dot.

1. INTRODUCTION

Semiconductor quantum dots being nanometer-scale islands have unique optical properties [1] that make them attractive candidates for many optoelectronic applications such as new types of single-photon sources [2, 3] and lasers [4, 5], or for use as computational building blocks of a quantum computer [6–8]. However, quantum dots are embedded in a surrounding solid, and the carriers confined to the dot interact with their environment most notably with phonons. Because of a strong suppression of phonon induced transitions between electronic states localized in quantum dots, the carrier-phonon interaction results in the decoherence of the optical polarization i.e. in pure dephasing being a major source for the decoherence. For S-shell excitons, the dominant sources of the pure dephasing are the longitudinal acoustic phonons [9–17]. Pure dephasing is caused by real irreducible processes. However, a significant contribution to the dot-reservoir interaction comes also from quantum fluctuations in which the reservoir degrees of freedom manifest themselves in a virtual state (i.e. the energies of the intermediate states are not equal to the energy of the initial and final state). Quantum fluctuations give rise to the shift of the dot energy levels known as the polaron shift. They also give a dominant contribution to self-energy functions of the quantum dots (QD) states. The energy dependence of QD self-energy functions can have a significant effect on the emission spectra of a strongly coupled quantum-dot cavity systems [18–21]. The problem is that despite the self-energy functions are used for description of the evolution QD interacting with environment, usually they are derived from quantum-field Green functions. A consistent description of quantum dynamics is ensured only by making use of the Green operator. In atomic physics where the self-energy functions of atomic states determine the Lamb shifts and widths of spectral lines the Green-function method has turned out to be very effective. The reason for this is the weakness of the electromagnetic interaction. However, the self-interaction of QD excitons is much more significant than the interaction of an atom with its own radiation field. In this paper we investigate the effect of

IOP Conf. Series: Journal of Physics: Conf. Series **859** (2017) 012013 doi:10.1088/1742-6596/859/1/012013

quantum fluctuations in a system consisting of a QD and the reservoir of acoustic phonons on processes in which the QD takes part. We find a contribution to the QD exciton self-energy function from the quantum fluctuations. In our investigation we make use of a generalized dynamical equation that has been derived by Gainutdinov [22] as the most general equation of motion consistent with the current concepts of quantum physics. We show that this equation that has turned out an important tool for solving many problems in nuclear physics [23–25], and quantum optics [26] provides a straightforward and effective way to describe the interaction of QDs with their environment.

2. Method

Being equivalent to the Schrödinger equation in the case of instantaneous interactions, the Gainutdinov equation of motion permits the generalization to the case where dynamics of a quantum system is governed by a nonlocal-in-time interaction. One of the energy representations [27–31] of this equation takes the form

$$\frac{dT(z)}{dz} = -T(z)(G_0(z))^2 T(z), \tag{1}$$

where $G_0(z) = (z - H_0)^{-1}$ with H_0 being the free Hamiltonian, and $T(z)$ is defined as

$$T(z) = i\int_0^\infty d\tau \exp(i(z - H_0)t_2)\tilde{S}(t_2, t_1)\exp(-i(z - H_0)t_1). \tag{2}$$

Here $\tilde{S}(t_2, t_1)$ is the contribution to the evolution operator from the process in which the interaction in the system begins at time t_1 and ends at time t_2. The boundary condition for equation (1) is,

$$T(z) \underset{|z|\to\infty}{\to} B(z) \tag{3}$$

where

$$B(z) = i\int_0^\infty d\tau \exp(iz\tau) H_{int}^{(s)}(\tau),$$

where

$$H_{int}^{(s)}(t_2 - t_1) = \exp(-i(z - H_0)t_2) H_I(t_2, t_1)\exp(-i(z - H_0)t_1),$$

with $H_I(t_2, t_1)$ being the generalized interaction operator.

The operator $T(z)$ determines the Green operator $G(z)$ being the evolution operator in the energy representation

$$G(z) = G_0(z) + G_0(z)T(z)G_0(z). \tag{4}$$

Equation of motion (1) with the boundary condition determines the dynamics of the system. The contribution to the Green operator $G(z)$, which comes from the processes associated with the self-interaction of the system, has the same structure as the free Green operator $G_0(z)$. For this reason it is natural to replace $G_0(z)$ by the operator $\tilde{G}_0(z)$, which describes the evolution of the system in the case where the interaction in the system is reduced to the self-interaction and hence has the structure

$$\tilde{G}_0(z) = (z - H_0 - C(z))^{-1}, \tag{5}$$

where the operator $C(z)$ has the same eigenvectors as H_0 ($C(z)|n\rangle = C_n(z)|n\rangle, H_0|n\rangle = E_n|n\rangle$). Correspondingly, the operator $T(z)$ is the operator describing the whole interaction in the system except the self-interaction. These operators are related as follows

$$G(z) = G_0(z) + G_0(z)T(z)G_0(z) = \tilde{G}_0(z) + \tilde{G}_0(z)M(z)\tilde{G}_0(z). \tag{6}$$

By making use of this equation the equation of motion (1) can be represented in the form of two equations for $M(z)$ and $C(z)$. Equation for the self-energy function takes the form

$$\frac{dC_n(z)}{dz} = -\sum_m \frac{\langle n|M(z)|m\rangle\langle m|M(z)|n\rangle}{(z - E_m - C_m(z))^2}. \tag{7}$$

3. Quantum fluctuations in phonons and the exciton self-energy function.

Let us consider the self-energy of a QD exciton within independent boson model (IBM) [18]. The IBM Hamiltonian describing phonons and exciton-phonon coupling reads [19]

$$\widehat{H}_{ph} + \widehat{H}_{ep} = \sum_q \omega_q b_q^* b_q + \sum_q g_x^q \left(b_q + b_q^+\right) |x\rangle\langle x|, \tag{8}$$

where $|x\rangle$ is the vector of the excitonic state, q denote the different phonon modes with energy ω_q, the creation (b_q^+) and annihilation (b_q) operators of phonons with momentum $\mathbf{q}$ and frequency ω_a obey the usual commutation relations for bosons, and G_x^q is the deformation potential coupling, which depends on the material parameters of the host semiconductor and the exciton wave function.

The self-energy function not only determines the energy shift of the exciton caused by the vacuum fluctuations in the QD exciton-phonon reservoir system, but also its energy dependence has a significant effect on the exciton spectrum. It has been shown [20] that a good account of the exciton spectrum can be obtained already at the second-Born perturbation level. At this order the solution of equation (7) takes the form

$$\frac{d\langle x,\nu|C^{(2)}(z)|x,\nu\rangle}{dz} = -\sum_\mu \frac{\langle x,\nu|H_I(z)|x,\mu >< x,\mu|H_I(z)|x,\nu\rangle}{(z - E_\mu)^2}. \tag{9}$$

By taking the thermal average of (9) we get

$$\frac{d\langle x,|C^{(2)}(E)|x\rangle}{dz} = -\sum_q \left\{\frac{|g(q)|^2(1+n(q))}{(E - E_x - w(q) + i0)^2} + \frac{|g(q)|^2 n(q)}{(E - E_x + w(q) - i0)^2}\right\}, \tag{10}$$

with $n(q) = [e^{\omega(q)/k_BT} - 1]^{-1}$ being the mean phonon occupation number at a bath temperature T. Solving this equation yields

$$C_x^{(2)}(E) = \sum_q \left\{\frac{|g(q)|^2(1+n(q))}{E - E_x - w(q) + i0} + \frac{|g(q)|^2 n(q)}{E - E_x - w(q) - i0}\right\}. \tag{11}$$

This equation can be rewritten as

$$C_i(E) = \frac{S_{HR}}{\omega_b^2} \int \omega^3 \exp\left(-\frac{\omega^2}{2\omega_b}\right) \left(\frac{1+n(q)}{E - E_i - \omega + i0} + \frac{n(q)}{E - E_i + \omega - i0}\right) d\omega, \tag{12}$$

where S_{HR} is dimensionless Huang-Rhys parameter characterizing the phonon-exciton coupling.

In general the self-energy function $C_x(E)$ can be represented as a sum of the contribution $\Sigma_x(E)$ from the off-shell quantum-fluctuation processes and the contribution $C_x^{on}(E)$ from on-shell processes leading to the system damping

$$C_x(E) = \Sigma_x(E) + C_x^{(on)}(E).$$

The interaction of the QD exciton with acoustic phonon reservoir generates pure quantum fluctuations that do not lead to decay processes, and, as a result, $C_x^{(2)}(E)$ gives contribution only to $\Sigma_x(E)$. Up to now we did not account for the self-energy processes caused by the interaction of the exciton with its own radiation field. These processes also give contributions to the quantum fluctuation function $\Sigma_x(E)$. However, this contribution is small compared with that from the quantum fluctuations in the QD exciton-phonon reservoir system and can be neglected. At the same time, the interaction of the exciton with the radiation field leads to its radiative recombination processes that being on-shell ones give contribution to the function $C^{(on)}(E) = -i\Gamma_r/2$ with Γ_r being the radiative decay rate.

For the effect of the quantum fluctuation to be strong the Huang-Rhys parameter S_{HR} must be large enough. In a field of exciton-acoustic phonon coupling (EPC) in QDs experimental studies generally show little agreement with theory or each other even among studies on the same material [32–34]. And it is hard to predict exciton-phonon coupling strength dependence on the nanocrystal radius or temperature [33, 34]. There are several experimental techniques that can be used to obtain EPC. Each method suffers from its own set of technical and/or interpretive complexities. The most straightforward is vibrationally resolved absorption or emission spectroscopy. In the low-temperature limit, the ratio of the one-phonon to zero-phonon intensity is equal to the parameter S_{HR} for that phonon mode. In Ref. [35] by using photoluminescence (PL) spectroscopy authors measured the Huang-Rhys constant for the system with InAs self-assembled quantum dots embedded in GaAs, and find the value $S_{HR} = 0.5$. They consider the Frohlich interaction of the electron-hole pair with bulk GaAs phonons in the system and interpret the phonon-assisted photoluminescence as being indicative of an enhanced electron-phonon interaction. The second type of experiments, resonance Raman scattering (RRS), provides the same type of information. Using this method it is much easier to measure quantities, depending on the Huang-Rhys parameter, but the interpretation of the data is less straightforward. In Ref. [33] authors obtain the Huang-Rhys parameter $S_{HR} = 0.7$ for PbS nanocrystals, 1.5 nm radius using RRS. Also, there are some domain measurements: femtosecond pump-probe, photon echo, etc. Short light pulses in these methods can impulsively excite those phonons that are coupled to the electronic excitation, producing a modulation in the signal as a function of probe delay. Here the Huang-Rhys factor can be obtained by Fourier transformation, followed by modelling of the time-dependent signal [34]. Experimental measurements, made in Ref. [32] show Huang-Rhys factor in range 0.15-0.35.

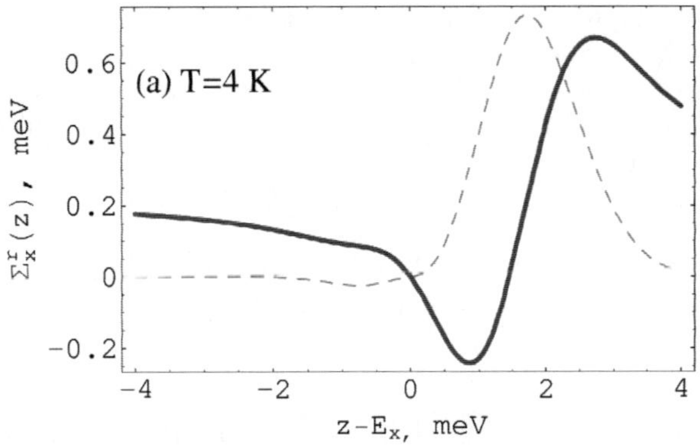

Figure 1. Quantum fluctuation contribution to the excitonic self-energy function (T = 4 K and S_{HR} = 0.2), where the solid curve represents the real contribution and the dashed curve the imaginary contribution.

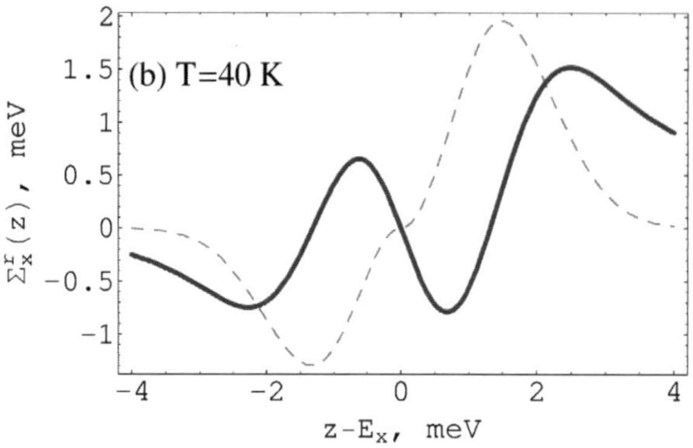

Figure 2. Quantum fluctuation contribution to the excitonic self-energy function ($T = 40$ K and $S_{HR} = 0.2$), where the solid curve represents the real contribution and the dashed curve the imaginary contribution.

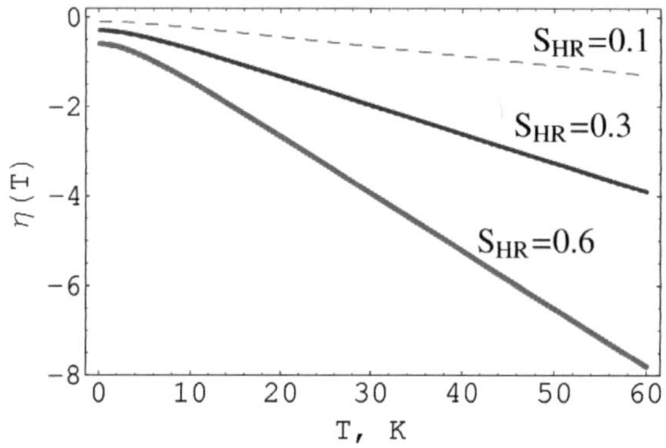

Figure 3. The temperature dependence of the parameter η.

The results of the calculations of the pure self-energy function presented in figures 1–3 show that the second-Born approximation we have made in deriving equation (11) actually is not well-defined for $S_{HR} \geq 0.1$ even at very low temperatures. In fact, in this case the self-energy function in the denominator of equation (7) cannot be neglected. Substituting $C_i(E)$ represented in the form

$$C_x(E) = (E - E_x)\eta + \tilde{C}_x(E), \tag{13}$$

(E_x is assumed to include the self-energy shift $C_x(E_x)$) with

$$\eta = \frac{dC_x(E)}{dE}|_{E=E_x} \tag{14}$$

into equation (11), and neglecting $\tilde{C}_x(E)$ that is small for the relevant energies we get

$$C_i^{(2)}(E) = \sum_q \left\{ \frac{|g(q)|^2 Z_1^2 (1 + n(q))}{(E - E_x - w(q) + i0)Z_2^2} + \frac{|g(q)|^2 Z_1^2 n(q)}{(E - E_x + w(q) - i0)Z_2^2} \right\}, \tag{15}$$

where $Z_2 = 1 - \eta$. The factor Z_2 can be regarded as a constant renormalizing the exciton propagator. Appearance of the factor Z_1 manifests the fact that the renormalization of the exciton propagator must be accompanied by a renormalization of the exciton-phonon coupling. An additional factor Z_2 appears because of the renormalization of the propagators associated with external lines. Since as usual $Z_1 = Z_2$, these constants in equation (15) compensate each other and we come back to equation (11), in which the parameter Z_2 does not manifest itself. However, there are physical situations where this parameter and hence the self-energy function $\Sigma_x(E)$ come into play. This, for example, takes place in the case when the quantum dot is strongly coupled to a cavity. In this case the cavity-emitted spectrum depends on the exciton self-energy function [18–21].

4. Conclusion

We have shown that the exciton self-interaction associated with quantum fluctuations in the combined system consisting of a semiconductor quantum dot and a reservoir of acoustic phonons can be strong. Our investigations show that this takes place in the case when Huang-Rhys parameter S_{HR} is more than 0.1. In this case the parameter Z_2 that determines strength of self interactions becomes more than one even at low temperatures. This parameter that in the renormalization theory renormalizes propagators of particles, in the case of semiconductor QDs can manifests itself, for example, in the cavity-emitted spectrum when a dot is embedded in a resonant cavity. The quantum fluctuation processes can manifest themselves not only in the self-energy function. In the case when the QD is coupled to resonance laser field they can give rise to the nonradiative transitions between the dressed states. A deeper understanding of these processes can provide a novel insight into the problem of the pure dephasing.

References

[1] Woggon U 1996 *Optical Properties of Semiconductors Quantum Dots* (Springer, Berlin) p 43
[2] Song H-Z, Hadi M, Zheng Y, Shen B, Zhang L, Ren Z, Gao R and Wang Z M 2017 *Nanoscale Research Letters* **12** 128
[3] Zhang J, Huo Y, Rastelli A, Zopf M, Höfer B, Chen Y, Ding F and Schmidt O G 2015 *Nano Letters* **15** 422-427
[4] Yadav A, Bai L, Yang Y, Liu J, Kaushik A, Cheng G J, Jiang L, Chi L and Kang Z 2017 *Nanoscale* **9** 5049-5054
[5] Wang C, Zhuang J-P, Grillot F and Chan S-C 2016 *Optics Express* **24** 29872-29880
[6] Delteil A, Sun Z, Fält S and Imamoğlu A 2017 *Phys. Rev. Lett.* **118** 177401
[7] Wei H-R and Deng F-G 2016 *Journal of the Optical Society of America* B **33** 804-809
[8] Li T, Yang and G-J Deng F-G 2016 *Phys. Rev.* A **93** 012302
[9] Gauger E M, Benjamin S C, Nazir A and Lovett B W 2008 *Phys. Rev.* B **77** 115322
[10] Ramsay A J, Achanta V G, Gauger E M, Nazir A, Lovett B W, Fox A M and Skolnick M S 2010 *Phys. Rev. Lett.* **104** 017402
[11] Förstner J, Weber C, Danckwerts J and Knorr A 2003 *Phys. Rev. Lett.* **91** 127401
[12] Machnikowski P and Jacak L 2004 *Phys. Rev.* B **69** 193302
[13] Vagov A, Croitoru M D, Axt V M, Kuhn T and Peeters F M 2007 *Phys. Rev. Lett.* **98** 227403
[14] Nazir A 2008 *Phys. Rev.* B **78** 153309
[15] Axt V M, Machnikowski P and Kuhn T 2005 *Phys. Rev.* B **71** 155305
[16] Hodgson T E, Viola L and DAmico I 2008 *Phys. Rev.* B **78** 165311
[17] Hohenester U and Stadler G 2004 *Phys. Rev. Lett.* **92** 196801
[18] Hughes S et al 2011 *Phys. Rev.* B **83** 165313
[19] Roy C and Hughes S 2011 *Phys. Rev.* X **1** 021009
[20] Tarel G and Savona V 2010 *Phys. Rev.* B **81** 075305
[21] Milde F, Knorr A and Hughes S 2008 *Phys. Rev.* B **78** 035330
[22] Gainutdinov R Kh 1999 *J. Phys.* A **32** 5657
[23] Gainutdinov R Kh and Mutygullina A A 2002 *Phys. Rev.* C **66** 014006
[24] Gainutdinov R Kh and Moutygoullina A A 1997 *Physics of Atomic Nuclei* **60** 841-847
[25] Gainutdinov R Kh and Mutygullina A A 2002 *Physics of Atomic Nuclei* **65** 1421-1430
[26] Gainutdinov R Kh and Mutygullina A A 2009 *Physica Scripta* T **135** 014012
[27] Gainutdinov R Kh 1989 *J. Phys.* A **22** 269-286
[28] Gainutdinov R Kh 1983 *Sov. J. Nucl. Phys.* **37** 277-282

[29] Gainutdinov R Kh 1987 *Sov. J. Nucl. Phys.* **46** 743-749
[30] Gainutdinov R Kh 1987 *Sov. J. Nucl. Phys.* **53** 885-892
[31] Gainutdinov R Kh 1995 *Zh. Eksp. Teor. Fiz.* **108** 1600-1613
[32] Sagar D M, Cooney R R, Sewall S L, Dias E A, Barsan M M, Butler I S and Kambhampati P 2008 *Phys. Rev.* B **77** 235321
[33] Krauss T D and Wise F W 1997 *Phys. Rev.* B **55** 9860
[34] Kelley A M 2010 *J. Phys. Chem. Lett.* **1** 1296-1300
[35] Förstner J, Weber C, Danckwerts J and Knorr A 2003 *Phys. Rev. Lett.* **91** 127401 García-Cristóbal A, Minnaert A W E, Fomin V M, Devreese J T, Silov A Yu, Haverkort J E M and Wolter J H 1999 *Phys. Stat. Sol.* B **215** 331-336.

Correlated signals of the accumulated photon echo on CdSe – CdS quantum dots

V V Samartsev[1,2] **and T G Mitrofanova**[1]

[1]E K Zavoisky Kazan Physical-Technical Institute, Sibirsky Trakt str. 10, Kazan 420029, Russia
[2]Kazan Federal University, Kremlevskaya str. 16, Kazan 420008, Russia

E-mail: samartsev@kfti.knc.ru

Abstract. The possibility of an observation of correlated signals of the accumulated photon echo under a three-photon excitation of a sample, which is an ensemble of semiconductor nanoparticles of "core – shell" type, by a femtosecond radiation of Ti:Sa laser is discussed. The phase matching conditions of such signals are obtained.

1. Introduction

Earlier, we predicted a possibility to observe the correlated free induction decay (CFID) signals [1]. Then, in paper [2], it was shown that the signals of stimulated photon echo (SPE) can also be correlated since they are in fact the same signals of free induction decay formed in a nonreversible system. Now let us pay attention to the fact that the correlation can exist also between the signals of accumulated photon echo (APE). The order of their excitation is similar to that of SPE, but at the presence of cumulative effect [3, 4]. The aim of this paper is to analyze the conditions of formation of the correlated APE responses on quantum dots CdSe – CdS. The peculiarities of a such quantum dots synthesis were presented in [5].

The correlation between the SPE signals as well as the correlation between CFID signals in [1] is understood as their identity which is due to the presence of EPR pairs (Einstein-Podolsky-Rosen pairs [6]) formed in the sample under its multi-photon excitation by two crossed beams of Ti:Sa laser radiation. The exciting beams are of the same 800-nm wavelength but they have different wave vectors $\mathbf{k}_1$ and $\mathbf{k}_2$. In the case of correlated FID signals two different nonequilibrium EPR-gratings $\mathbf{k}_1$ - $\mathbf{k}_2$ and $\mathbf{k}_2$ - $\mathbf{k}_1$ simultaneously appear in a resonant medium as a result of excitation. The correlated signals are in fact the coherent responses scattered by these gratings in opposite directions. As for both correlated SPE and APE signals the situation is more complicated. There are two exciting pulses which act at one and the same area of a sample before the influence of the third (reading) pulse $\mathbf{k}_3$. It is well-known fact that SPE as well as APE signals are formed due to the diagonal part of a density matrix after two exciting pulses. That is why in this case EPR-gratings $\mathbf{k}_1$ - $\mathbf{k}_2$ and $\mathbf{k}_2$ - $\mathbf{k}_1$ are the nonequilibrium population gratings formed after the action of two exciting pulses. The theory of EPR-gratings formation in nanostructured semiconductor sample under its multi-photon excitation by laser radiation was presented in [7]. The special case of the sample containing CdSe–CdS colloidal quantum dots was discussed in [2].

Now we should say a few words about the mechanism connecting the electrons in EPR-pairs in a semiconductor sample. In the experimental study [8] it was established that in the high density gas of carriers, which appears in CdS crystal under two-photon mode of its excitation, an essential role belongs to excitons and excitonic molecules. So, the EPR-pairs are seemed to be biexcitons. Since such femtosecond experiments are performed at room temperature, the phonons play the important role also. The situation is quite similar to Cooper pairs in superconductors when two electrons with opposite spins form a bound state through the exchange of phonons [9]. In our case the mechanism of EPR-pairs formation is more complicated since one should take into account the Coloumb screening [10].

2. Peculiarities of CdSe–CdS nanopowders

The development of nanotechnologies allows us to create and to investigate the nanodimensional objects with unique optical properties [9]. For example, there is such quantum size effect as the dependence of a forbidden band width on a nanoparticle size, which provides a possibility to control a wavelength of luminescence. Such nanomaterials can be used for a creation of optoelectronic emitters, quantum dots displays and new lasers.

Earlier, in [5], we have already reported about synthesis and study of semiconductor nanoparticles of "core – shell" type, where the CdSe nanoparticles play the role of a core covered by the CdS shell. The size of CdSe core was equal to 3–4 nm and related to the thickness of the shell as 2,5:1. It was found that in the presence of the CdS shell the luminescence caused by the femtosecond 800-nm Ti:Sa laser radiation increased in 8 times by contrast with the pure core and was due to the processes of both two- and three-photon absorption. The luminescence efficiency increases due to the specific configuration of energy zones, which leads to more effective excitation of the sample. A smaller number of internal elementary excitations decays nonradiatively. The band structure and the scheme of energy transitions between bands as well as the luminescence spectra of CdSe–CdS nanocomposites are shown in figure 1.

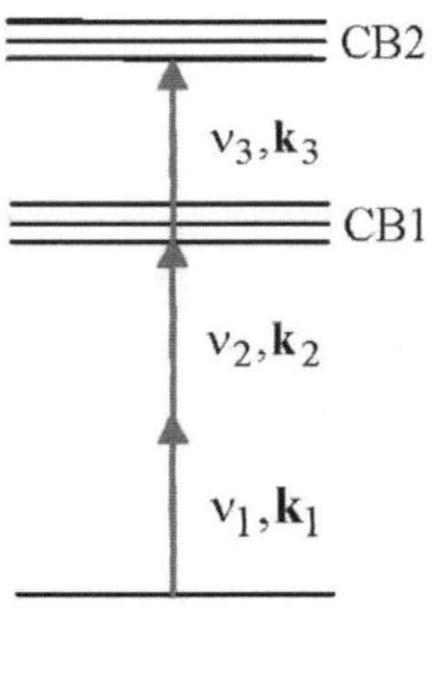

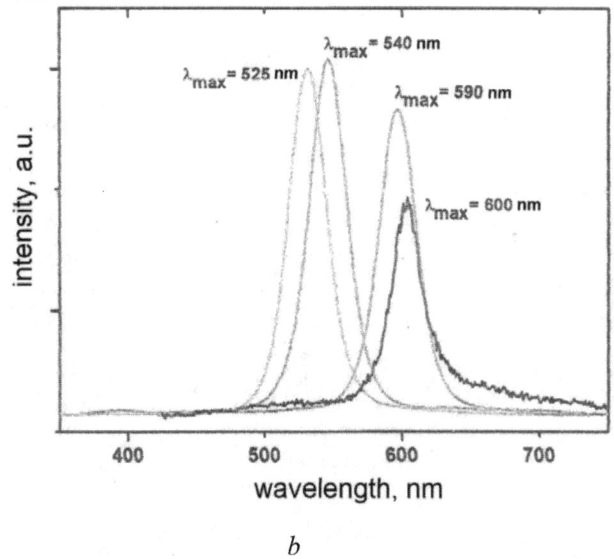

a *b*

Figure 1. Band structure (a) and luminescence spectra (b) of CdSe–CdS nanoparticles, CB1 and CB2 are the conduction bands, ν_i and $\mathbf{k}_i$ are the frequency and the wave vector of exciting pulses, i=1, 2, 3.

IOP Conf. Series: Journal of Physics: Conf. Series **859** (2017) 012014 doi:10.1088/1742-6596/859/1/012014

One can see that for the particles of 3 nm size the emission maximum occurs at 540 nm, of 6,5 nm size – at 590 nm and of the 7 nm size – at 600 nm. So, we observe the luminescence from the conductive band CB1 while the excitation was performed to the conductive band CB2, i.e. the nonradiative transition CB1→CB2 takes place. One may suppose that the nonequilibrium gratings $\mathbf{k}_1$ - $\mathbf{k}_2$ and $\mathbf{k}_2$ - $\mathbf{k}_1$ should disappear in this case. Nevertheless, it was shown by the authors of experimental study [10] and in our own experiments that with the three-photon excitation of the sample at 800 nm the population gratings exist both at 267 nm and 534 nm transitions. Due to the nonradiative relaxation the population grating at 267 nm transition disappears but it still exist at 534 nm transition and scatters coherently the correlated APE responses.

3. Correlated APE signals on CdSe–CdS nanoparticels

The order of APE excitation is well known [3, 11]. The sample is firstly acted by *n* pairs of exciting pulses with wave vectors $\mathbf{k}_1$ and $\mathbf{k}_2$ and then the APE signal is red out by means of the separate pulse with the wave vector $\mathbf{k}_3$. The APE signal is in fact the co-phased set of *n* SPE signals. In this paper we suggest to use the four-wave mixing technique for SPE generation and, namely the experimental scheme presented in [12], but instead of single exciting pulses ($\mathbf{k}_1$ and $\mathbf{k}_2$) we shall consider the periodic two-pulse train. One of the convenient regimes of an excitation is realized when the exciting pulse $\mathbf{k}_1$ acts at the angle of 30° to the axis of the sample, the exciting pulse $\mathbf{k}_2$ – at the angle of 150° and the readout pulse $\mathbf{k}_3$ – at the angle of 90°. In this case one of the expected correlated APE signals will be directed along $\mathbf{k}_2$, the second one – along $\mathbf{k}_1$ (both signals has the carrier wavelength equal to 267 nm × 2 = 534 nm which differs from the carrier wavelength of the exciting pulses) and the third one will not be emitted. Since each of exciting pulses should propagate in its own direction we have three channels (1, 2, 3) illustrated in figure 2.

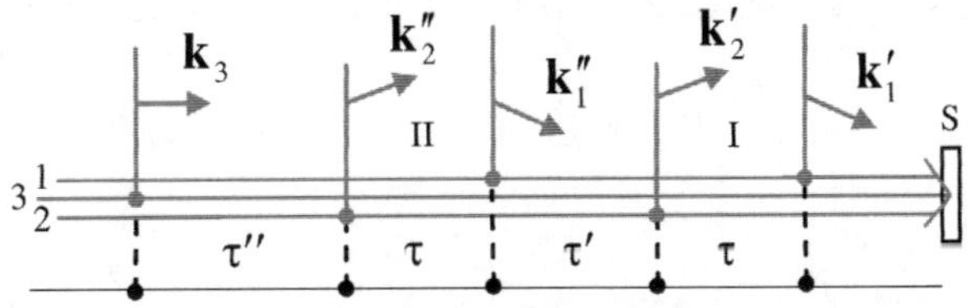

Figure 2. Three-channel (1, 2, 3) regime of an excitation of the correlated APE signals. I and II are two pairs of exciting pulses with wave vectors $\mathbf{k}_1$ and $\mathbf{k}_2$, $\mathbf{k}_3$ is a wave vector of a readout pulse, τ, τ′, τ″ are time intervals, S is a sample.

The corresponding scheme of a multipulsed excitation of APE signal similar to that from [12] is presented in figure 3. The optical regenerative amplifier ORA1 participates in the formation of pulses with wave vectors $\mathbf{k}_1'$ и $\mathbf{k}_1''$ and also of the readout pulse $\mathbf{k}_3$ while the optical regenerative amplifier ORA2 serves to the formation of pulses with wave vectors $\mathbf{k}_2'$ и $\mathbf{k}_2''$. As a result two pairs of exciting pulses with a certain time delay between them are produced.

For such mutual spatial arrangement of the wave vectors of exciting pulses the phase matching conditions of correlated APE signals can be written as

$$\mathbf{k}_4 = -\mathbf{k}_1 + \mathbf{k}_2 + \mathbf{k}_3, \quad (1)$$
$$\mathbf{k}_5 = \mathbf{k}_1 - \mathbf{k}_2 + \mathbf{k}_3, \quad (2)$$
$$\mathbf{k}_6 = \mathbf{k}_1 + \mathbf{k}_2 - \mathbf{k}_3 = 0. \quad (3)$$

The situation is illustrated in figure 4.

IOP Conf. Series: Journal of Physics: Conf. Series **859** (2017) 012014 doi:10.1088/1742-6596/859/1/012014

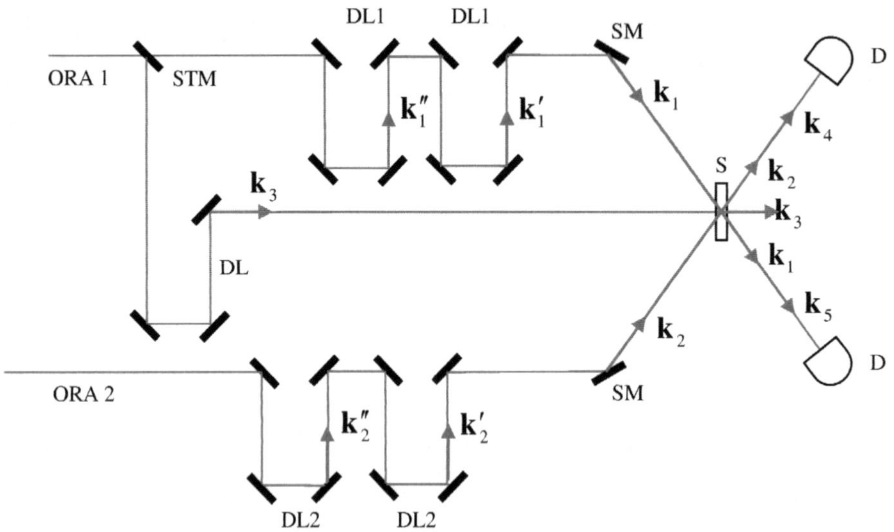

Figure 3. Experimental scheme of an excitation of the correlated APE signals on nanostructured quantum dots CdSe–CdS. $\mathbf{k}'_1$, $\mathbf{k}''_1$, $\mathbf{k}'_2$, $\mathbf{k}''_2$, $\mathbf{k}_3$ are the wave vectors of exciting pulses, $\mathbf{k}_4$ and $\mathbf{k}_5$ are the wave vectors of the correlated APE signals, DL is a delay line, D is a detector, S is a sample, STM is a semitransparent mirror, SM is a spherical mirror, ORA1 and ORA2 are optical regenerative amplifiers.

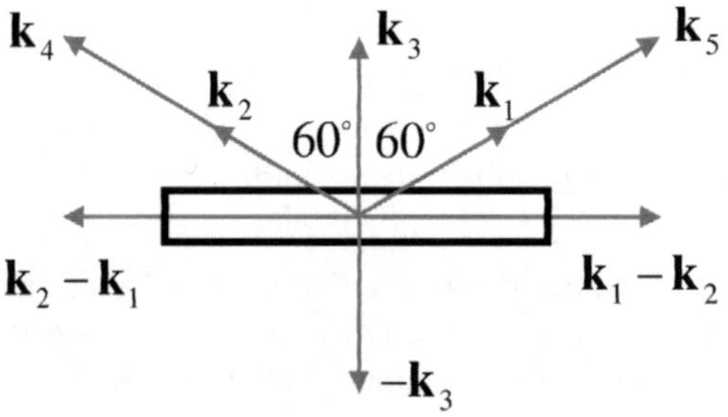

Figure 4. Phase matching conditions of the correlated APE signals ($\mathbf{k}_4$ and $\mathbf{k}_5$) under a three-photon excitation of the nanostructured CdSe–CdS sample at 800 nm. The carrier wavelength of the correlated APE signals is equal to 534 nm.

Let us briefly discuss the method of calculation of the correlated APE signals. Following [13] we write the intensity of optical coherent responses emitted into the solid angle $\Delta\Omega$ as

$$\Delta I(\mathbf{k}_{res}, t) = I_0(\mathbf{k}_{res}) F(t) F^*(t) \Delta\Omega, \tag{4}$$

where $I_0(\mathbf{k}_{res})$ is the intensity of a three-quantum spontaneous radiation of an isolated electron in the direction of $\mathbf{k}_{res}$,

$$F(t) = 2\int_0^\infty d(\Delta\omega) g(\Delta\omega) \sum_j^N W(\Delta\omega, \mathbf{r}_j, t) e^{-i\mathbf{k}_{res}\mathbf{r}_j} , \qquad (5)$$

$\Delta\omega = \omega_e - 3\omega_L$ is a detuning parameter, $g(\Delta\omega)$ is a distribution function of $\Delta\omega$, $W(\Delta\omega, \mathbf{r}_j, t)$ – an element of a density matrix of an electron system, corresponding to a coherent response. The electrical field of the correlated APE signal ($\mathbf{k}_4$) is equal to [11]:

$$\mathbf{E}_{\mathrm{CAPE}} = -N\mathbf{E}_0(R)\frac{1}{2}\theta_1\theta_2\theta_3 G e^{-i\Delta\omega(t-3\tau-\tau'-\tau'')} e^{-i(-\mathbf{k}_1+\mathbf{k}_2+\mathbf{k}_3-\mathbf{k}_{CAPE})\mathbf{r}} e^{-\frac{2\tau}{T_2}-\frac{\tau'}{T_1}} \cdot \left[\frac{1-\exp(-n\tau'/T_1)}{1-\exp(-\tau'/T_1)}\right], \qquad (6)$$

where N is a number of nanoparticles, θ_η is an area of η-th pulse, τ, τ′, τ′′ are the time intervals illustrated in figure 2, n is a number of pulse pairs, G is a parameter. It is worth mentioning that all calculations are provided using the approximation of small pulse areas $\sin\theta_\alpha \approx \theta_\alpha$. We suppose that, despite the smallness of a three-photon excitation cross-section (10^{-79} cm^6s^2photon number^{-2}), these coherent responses are still observable [10]. The possibility to observe the correlated APE signals was also noted in [14].

Acknowledgments
This study was supported by the Program of the Presidium of RAS "Actual problems of physics of low temperatures" and RFBR grant no. 17-02-00701-a.

References

[1] Leontiev A V, Lobkov V S, Mitrofanova T G, Samartsev V V and Shmelev A G 2012 *Laser Phys. Lett.* **9** 654
[2] Samartsev V V and Mitrofanova T G 2016 *Laser Phys.* **26** 125203
[3] Yevseyev I V, Rubtsova N N and Samartsev V V 2009 *Coherent Transient Processes in Optics* (Moscow: Fizmatlit) (in Russian)
[4] Zuikov V A, Samartsev V V, Stelmakh M F, Yakshin M A, Yufin M A and Yashin A N 1991 *Laser Phys.* **1** 678
[5] Zharkov D K, Shmelev A G, Leontiev A V, Lobkov V S and Samartsev V V 2014 *Coherent Optics and Optical Spectroscopy* ed M Kh Salakhov (Kazan: Kazan Federal University) pp. 109–112
[6] Einstein A, Podolsky B and Rosen N 1935 *Phys. Rev.* **47** 777
[7] Samartsev V V 2015 *Correlated Photons and Their Applications* (Cambridge, UK: CISP)
[8] Hillmann F, Voight J and Redlin H 2000 *Appl. Phys. Lett.* **77** 4181
[9] Roco M C, Williams R S and Alivisatos P (Eds) 2000 *Nanotechnology Research Directions: IWGN Workshop Report: Vision for Nanotechnology R & D in the Next Decade* (Dordrecht: Kluwer Acad. Publ.)
[10] Chon J W M and Gu M 2004 *Appl. Phys. Lett.* **84** 4472
[11] Kalachev A A and Samartsev V V 2003 *Coherent Phenomena in Optics* (Kazan: Kazan University) (in Russian)
[12] Mckimmie L J, Lincoln C N, Jasieniak J and Smith T A 2010 *Jour. Phys. Chem.* **114** 82
[13] Manykin E A and Samartsev V V 1984 *Optical Echo Spectroscopy* (Moscow: Nauka) (in Russian)
[14] Takemoto K, Hyun B R and Masumoto Y 2000 *Jour. of Luminescence* **87-89** 485

IOP Conf. Series: Journal of Physics: Conf. Series **859** (2017) 012015 doi:10.1088/1742-6596/859/1/012015

On the approximations of slowly varying envelope and slowly varying profile in nonlinear optics

S V Sazonov

National Research Centre «Kurchatov Institute», 123182 Moscow, Russia

e-mail: sazonov.sergey@gmail.com

Abstract. The comparative analysis of two specified in heading approximations is carried out. It is shown that both approximations mutually supplement each other, finding applications in nonlinear optics of pulse duration from nano- to femtoseconds.

1. Introduction

Since the 60s of the last century one began to create pulse lasers in laboratory conditions. Pulses of nanosecond duration were generated at that time. There were at once questions concerning the theoretical studies of interaction of such pulses with matter. In recent years duration of the generated pulses has reached femtoseconds. Such signals contain about one period of electromagnetic oscillations. Therefore their spectrum range is very wide. This spectrum can contain Fourier-components, which are resonant to various quantum transitions. As the result, it can lead to strong excitations of media. The simplest models of media lead to difficult nonlinear self-consistent equations. The two-level model well describes the interaction of medium with resonant or quasi-resonant optical pulses. In this case the carrier pulse frequency ω is close to the frequency ω_0 of a quantum transition. To move ahead further, it is necessary to use various physically reasonable approximations. In the present article we will discuss two approximations.

From the first look at the heading of the present work, the difference between two designated approximations is not visible. There should not be confusion, as these approximations make absolutely different physical senses. The present work is devoted to the disclosure of the essence and determination of distinctions between these approximations.

2. Approximation of slowly varying envelope

Let the carrier frequency of an optical pulse belong to the visible or infrared spectral range ($\omega \sim 10^{15} c^{-1}$). Then the pulses whose duration τ_p lies in the range from nano - to picoseconds contain from one million to one thousand oscillations. If to designate the last number as N , then for such pulses it is possible to determine a small parameter

$$\mu_1 = N^{-1} \sim \left(\omega\tau_p\right)^{-1} << 1. \qquad (1)$$

The sense of condition (1) consists in the fact that here it is possible to define a concept of an envelope of a light pulse. This envelope is a function, which varies slowly at times of the order of the oscillation period and at distances about a wavelength. In this case the electric field of the pulse can be presented in the form

$$E(z,t) = \psi(z,t)e^{i(\omega t - kz)} + \psi^*(z,t)e^{-i(\omega t - kz)}, \qquad (2)$$

where z is the distance passed by the pulse, t is time, k is the wave number, corresponding to the carrier frequency ω, ψ is the complex pulse envelope, which satisfies the conditions

$$|\psi_t| << \omega|\psi|, \qquad |\psi_z| << k|\psi|. \tag{3}$$

Let us illustrate the application of the slowly varying envelope approximation (SVEA) on the example of interaction of a light pulse with the two-level medium. The corresponding self-consistent set of equations looks like [1]

$$E_{zz} - (n_m / c)^2 E_{tt} = (4\pi / c^2) P_{tt} = aU_{tt} = -a\omega_0 V_t, \tag{4}$$

$$U_t = -\omega_0 V, \qquad V_t = \omega_0 U + 2(dE/\hbar)W, \qquad W_t = -2(dE/\hbar)V. \tag{5}$$

Here c is the speed of light, $\hbar$ is the Planck constant, n_m is the refractive index of the matrix, $a = 8\pi dn / c^2$, n is the concentration of two-level atoms, d is the dipole moment of the considered quantum transition, the material variables U and W have the meaning of dimensionless induced atomic dipole moment and half the difference of population of the excited and basic quantum levels (population inversion), respectively, V is the growth rate of the change of the induced dipole moment.

Despite the simplicity of the model of the two-level medium, the set of equations (4) and (5) is very difficult. This set is nonlinear, and the coefficients in two last equations are quickly changing on times of the period of light oscillations. This set would be much simpler if the temporal scale of changes of dynamical variables increased up to the duration of light pulses. The SVEA allows to make this. Substituting (2) into (5) and neglecting the second derivatives of the complex envelope ψ, according to inequalities (3), we obtain approximately

$$E_{tt} = \left(2i\omega\psi_t - \omega^2\psi\right)e^{i(\omega t - kz)} + c.c., E_{zz} = \left(-2ik\psi_z - k^2\psi\right)e^{i(\omega t - kz)} + c.c. \tag{6}$$

Let us return to system of the material equations (5) now. Let us consider that $S = U + iV$. Then from the first two equations (5) we have

$$S_t = i\omega_0 S + 2i(dE/\hbar)W. \tag{7}$$

In the absence of the light pulse field ($E = 0$) this equation describes the movement of a free complex oscillator with eigen-frequency. In this case we have $S \sim \exp(i\omega_0 t)$. Obviously, it corresponds to the fact that the dipole moment of the excited two-level atom oscillates on the frequency of its quantum transition. The last term in the right-hand part can be considered as an external force the frequency of which is equal to the carrier frequency ω of the light pulse. It is clear, that steady state oscillations of the oscillator occur on this frequency. Therefore, we will write

$$S(z,t) = R(z,t)e^{i(\omega t - kz)}, \tag{8}$$

where $R(z,t)$ is the complex envelope of the atomic dipole moment.

Presenting the variable V in the form

$$V = (S - S^*)/2i = -i\left(\mathrm{R}\, e^{i(\omega t - kz)} - \mathrm{R}^*\, e^{-i(\omega t - kz)}\right)/2 \tag{9}$$

and neglecting the derivative of the envelope, we will have

$$V_t = \omega\left(Re^{i(\omega t - kz)} + R^* e^{-i(\omega t - kz)}\right)/2.$$

Substituting this expression, and also (2) and (6) into (4), – after equating expressions at $e^{i(\omega t - kz)}$ and $e^{-i(\omega t - kz)}$ in the left- and right-hand parts, we obtain

$$\psi_z + (n_m / c)\psi_t = -i\beta R, \tag{10}$$

where $\beta = 4\pi d^2 n\omega_0 / (\hbar c n_m)$.

By obtaining (10) we have equated to zero coefficients at the free terms ψ and ψ^*. It has allowed to find the dispersion equation $k = \omega n_m / c$.

Thus, the use of SVEA has allowed to reduce wave equation (4) from the second order to the first order with respect to the derivatives.

Substituting (8) and (2) into (7), we find $R_t = i(\omega_0 - \omega)R + i(\psi + \psi^* e^{-2i(\omega t - kz)})W$. Here it is possible to neglect the term oscillating on the frequency 2ω. Really, the characteristic temporal scale of change of ψ corresponds to a pulse duration τ_p that is, owing to (3), much larger than the period of oscillations of the imaginary exponent. Then with a good accuracy we have

$$R_t = i\Delta R + i\psi W, \tag{11}$$

where $\Delta = \omega_0 - \omega$ is the frequency detuning from resonance.

Substituting (2), (8) and (9) into third equation (5) and also neglecting the terms oscillating on the frequency 2ω, we obtain

$$W_t = i(\psi^* R - \psi R^*)/2. \tag{12}$$

Equations (10) – (12) are well-known and bear the name of the Maxwell — Bloch (MB) system. It was the first integrable system generating solutions in the form of optical solitons [2].

Let us consider a case when frequency detuning meets the quasi-resonance condition [3]

$$\mu_2 \equiv (\Delta\tau_p)^{-1} << 1. \tag{13}$$

In this case it is possible to exclude the material variables from system (10) – (12). For this purpose we rewrite (11) in the form

$$R = -\psi W / \Delta - iR_t / \Delta. \tag{14}$$

The ratio of the second term in the right-hand part of (14) to the left-hand part is $\sim \mu_2 << 1$. Therefore, it is possible to use the method of consecutive approximations with respect to the second term in the right-hand part of (14). In the zero-order approximation we have $R = -\psi W / \Delta$. Substituting this expression into the mentioned small term, we will find as a first approximation $R = -(\psi / \Delta)W + i\Delta^{-2}(\psi W)_t$. Continuing the approximations, we will come to the Crisp expansion [4]

$$R = -(\psi / \Delta)W + i\Delta^{-2}(\psi W)_t + \Delta^{-3}(\psi W)_{tt}. \tag{15}$$

Let us take into consideration that the interaction of the pulse with the medium under quasi-resonance conditions is weak. Therefore, in the second and third terms of the right-hand part of (15) we will neglect the change of the difference of population densities, believing $W = W_\infty$. Then we have

$$R = -(\psi / \Delta)W + iW_\infty \Delta^{-2}\psi_t + W_\infty \Delta^{-3}\psi_{tt}. \tag{15a}$$

As the difference of the population densities of atomic quantum levels under quasi-resonance conditions changes slightly, we will substitute (15a) into (12), restricting ourselves to the first two terms of this expression. Then $W_t = -W_\infty(\psi^*\psi_t + \psi\psi_t^*)/2\Delta^2$. After integration of this equation we obtain

$$W = W_\infty(1 - |\psi|^2 / 2\Delta^2). \tag{16}$$

After substitution of (15a) and (16) into (10) we arrive at the Nonlinear Schrödinger (NLS) equation

$$i\Phi_z = -(k_2/2)\Phi_{\tau\tau} - b|\Phi|^2\Phi, \tag{17}$$

where $\Phi = \psi\exp(-i\beta W_\infty z/\Delta)$, $\tau = t - z/\upsilon_g$, $b = -\beta W_\infty/2\Delta^3$, $k_2 = \partial(v_g^{-1})/\partial\omega$ is the coefficient of group velocity dispersion, the linear group velocity υ_g is determined by the expression $1/\upsilon_g = n_m/c - \beta W_\infty/\Delta^2$.

Generally, NLS describes propagation of optical solitons in the isotropic non-resonant dielectrics having cubic (Kerr) nonlinearity [5, 6]. The model of the two-level medium accepted here is only an illustration of this fact.

3. Approximation of slowly varying profile

The pulse with the duration near 1 femtosecond may contain one or several periods of light oscillations. According to the terminology which has developed so far such signals are called the "few-cycle pulses" (FCP) [7]. In this case the parameter μ_1 stops being small and its value becomes unit order. It is clear, that in such conditions it is impossible to tell about pulse envelope. Here it is necessary to look for other methods and approaches. In 1973 in work [8] an alternative approach to SVEA has been offered. Authors of the mentioned work used the approach of the medium of small concentration of two-level atoms. This approximation corresponds to the inequality

$$\mu_3 \equiv 8\pi d^2 n/\hbar\omega_0 << 1. \tag{18}$$

The right-hand part of (4) is proportional to small parameter μ_3, and therefore it can be considered small. In such a situation it is possible to apply the slowly varying profile approximation (SVPA) the essence of which is revealed below. If to put the right-hand part of (4) equal to zero, then we have the well-known solution consisting of superposition of two waves. These waves propagate respectively along and against a z-axis with a velocity c/n_m. In approximation (18), the part of the pulse field $E(z,t)$ scattered back, against a z-axis, is negligible. Therefore, it is possible to consider that the pulse propagates only along the z-axis. That is, when the right-hand part equals zero, we have the solution $E = E(t - n_m z/c)$. This assumption allows to lower the order of wave equation (4). To account for the right-hand part of (4), we will determine the "local" time and "slow" coordinate: $\tau = t - n_m z/c$, $\zeta = \mu_3 z$. We assume that $E = E(\tau,\zeta)$. Then

$$E_z = \tau_z E_\tau + \zeta_z E_\zeta = -(n_m/c)E_\tau + \mu_3 E_\zeta, \qquad E_t = \tau_t E_\tau + \zeta_t E_\zeta = E_\tau.$$

Neglecting the small term, proportional to μ_3^2, we will write down

$$E_{zz} \approx (n_m^2/c^2)E_{\tau\tau} - 2\mu_3(n_m/c)E_{\tau\zeta}, \; E_{tt} = E_{\tau\tau}.$$

As a result, equation (4) after integration with respect to g and return to the initial independent variables will take the form

$$E_z + (n_m/c)E_t = gV. \tag{19}$$

Here $g = 4\pi dn\omega_0/cn_m$.

The procedure of reduction of equation (4) to look (19) is carried out in detail here in order that it is clear that this reduction has no slightest hint on that how many oscillations of the light field can be contained in the pulse. Then there can be any quantity. This is the advantage of SVPA in comparison with SVEA. On the other hand, it follows from the procedure of reduction of wave equation (4), which is carried out above, that the velocity of an optical pulse slightly differs from linear velocity. Therefore, under condition of SVPA it is impossible to describe considerable reduction of velocity of the pulse propagation.

The procedure which is carried out above allows to understand the sense of the discussed approximation. The dependence of the form $E = E(\tau, \zeta)$ reflects the fact of slow deformation of the profile of electric field in the fame of reference moving with a linear velocity c / n_m.

Let us notice that the term "the slowly varying profile approximation" has passed to optics from works on acoustics [9].

System (19), (5) is called the reduced Maxwell – Bloch (RMB) system. As was shown in [8], this system is integrable, possessing soliton solutions.

Further in various works (see, e.g. [10 – 14]) the refusal from the approximation of the medium of small atomic concentration has been made. The approximations of weak nonlinearity and (or) dispersion were used. Also the refusal from the model of the two-level medium has been made [15]. Various phenomenological models of media were offered [16].

Let us shortly pay attention to the medium with square nonlinearity and weak dispersion. The polarizing response of such medium can be presented as $P = P_{lin} + P_{non}$, where linear and nonlinear parts of the response have respectively the appearance $P_{lin} = \int_0^\infty \chi(t')E(t-t')dt'$, $P_{non} = \chi^{(2)}E^2$, $\chi(\tau)$ and $\chi^{(2)}$ are linear and nonlinear susceptibilities. Making expansion $E(t-t') = E(t) + t'^2 E_{tt}/2 - ...$ and substituting the obtained expression for P into the right-hand part of equation (4) at $n_m = 1$, after using SVPA we will come to the well-known Korteweg – de Vries equation:

$$E_z + qEE_\tau - \alpha E_{\tau\tau\tau} = 0. \tag{20}$$

Here $\tau = t - n_0 z / c$, $n_0 = \sqrt{1 + 4\pi\chi_0}$ is the low-frequency refractive index, $\chi_0 = \int_0^\infty \chi(\tau)d\tau$, $\chi_\omega = \int_0^\infty \chi(\tau)e^{-i\omega\tau}d\tau$ are the low-frequency and high-frequency susceptibilities, respectively, $q = 4\pi\chi^{(2)}/(cn_0)$, $\alpha = -(cn_0)^{-1}\int_0^\infty \tau^2\chi(\tau)d\tau = (cn_0)^{-1}\left(\partial^2\chi_\omega/\partial\omega^2\right)_{\omega=0}$ are the nonlinear and dispersion coefficients, respectively.

Various solutions of equation (20) are in detail discussed in monographs (see, e. g. [2]). Here it is more important to notice that on the basis of SVEA and SVPA it is possible to obtain the systems describing well-known effects in nonlinear optics. As an example we will consider the effect of optical rectification [17]. For this purpose we will present electric field in the form

$$E(z,t) = \Sigma + \psi(z,t)e^{i(\omega t - k_1 z)} + \psi(z,t)e^{-i(\omega t - k_1 z)}, \tag{21}$$

where Σ is the low-frequency pulse component, generated by means of optical rectification. Usually the spectrum of this component lies in the terahertz area. Therefore, we will call it a terahertz component. Let us neglect the own nonlinearity and dispersion of terahertz components. Then,

$$P_{non} = \left(2\chi^{(2)}\Sigma\psi e^{i(\omega t - kz)} + c.c.\right) + 2\chi^{(2)}|\psi|^2,$$

$$P_{lin} = \chi_\omega\psi e^{i(\omega t - kz)} - i\left(\partial\chi_\omega/\partial\omega\right)\psi_t - 0.5\left(\partial^2\chi_\omega/\partial\omega^2\right)\psi_{tt} + ... + c.c.$$

From here, from (21) and (4) after using SVPA we will obtain [15]

$$i\psi_z + (k_2/2)\psi_{\tau\tau} = \beta_1\Sigma\psi, \tag{22}$$

$$\Sigma_z = -\beta_2\left(|\psi|^2\right)_\tau, \tag{23}$$

where $k_2 = \partial(1/v_g)/\partial\omega = (4\pi/cn_\omega)\left[\partial\chi_\omega/\partial\omega + 0.5\omega\partial^2\chi_\omega/\partial\omega^2 - (\omega/4\pi)(\partial n_\omega/\partial\omega)^2\right]$, $n_\omega = \sqrt{1 + 4\pi\chi_\omega}$ is the refractive index corresponding to the frequency ω. Here we have put that the group velocity $v_g = c\left(n_\omega + \omega\partial n_\omega / \partial\omega\right)^{-1}$ of optical components is equal to the phase velocity $v_T = c / n_0$ of the terahertz component. In this case generation of terahertz components happens most intensively, $\beta_1=4\pi\chi^{(2)}\omega/(cn_0)$, $\beta_2=4\pi\chi^{(2)}/(cn_0)$.

System (22), (23) is known as the Yajima - Oikawa system [18]. The soliton solution of this system describes the process of an optical way of generation of terahertz radiation [19], finding now various applications [20].

Let us highlight that in system (22), (23) ψ is the envelope of the pulse electric field. At the same time, Σ is the electric field of a terahertz component. Thus, by derivation of system (22), (23) both approaches – SVEA and SVPA were used.

4. Conclusion

The analysis which is carried out above allows to draw a conclusion that both approaches (SVEA and SVPA) play an important role in nonlinear optics of light pulses. On the one hand, the approach of SVPA allows to describe propagation in the media of both quasi-monochromatic pulses, and FCP. At the same time by means of SVEA it is possible to consider only dynamics of quasi-monochromatic pulses. On the other hand, by means of the approach of SVEA it is possible to describe propagation of pulses in the media with the strong dispersion (linear and nonlinear) leading to considerable reduction of group velocity in comparison with linear velocity of light. The approach of SVPA is adapted only for cases when dispersion is small, and reduction of group velocity is insignificant.

Thus, it is possible to state that both approaches (SVEA and SVPA) are absolutely different in the physical sense. At the same time, both approaches mutually supplement each other at the studies of problems of the modern nonlinear optics.

Acknowledgment

This work is supported by the Russian Foundation for Basic Research (grant No. 16-02-00453-a).

References

[1] Allen L and Eberly J H 1978 *Optical resonance and two-level atoms* (New York: John Wiley & Sons)
[2] Lamb G L 1980 *Elements of soliton theory* (New York: John Wiley & Sons)
[3] Basharov A M and Maimistov A I 2000 *Optika i Spektroskopiya* **88** 428
[4] Crisp M D 1973 *Phys. Rev.* A **8** 2128
[5] Agrawal G P 1989 *Nonlinear fiber optics* (New York: Academic Press)
[6] Kivshar Yu S and Agrawal G P 2003 *Optical solitons* (New York: Academic Press)
[7] Brabec T and Krausz F 2000 *Rev. Mod. Phys.* **72** 545
[8] Eilbeck J C, Gibbon J D, Caudrey P J and Bullough R K 1973 *J. Phys.* A **6** 1337
[9] Akhmanov S A 1986 *Uspekhi Fizicheskikh Nauk* **49** 361
[10] Belenov E M, Nazarkin A V and Ushchapovskii V A 1991 *Sov. Phys. JETP* **73** 422
[11] Sazonov S V and Trifonov E V 1994 *J. Phys.* B **27** L7
[12] Leblond H, Sazonov S V, Mel'nikov I V, Mihalache D and Sanchez F 2006 *Phys. Rev.* A **74** 063815
[13] Maimistov A I 2000 *Quantum Electron* **30** 287
[14] Sazonov S V 2016 *Opt. Commun.* **380** 480
[15] Sazonov S V and Sobolevskii 2003 *JETP* **96** 1019
[16] Kozlov S A and Sazonov S V 1997 *JETP* **84** 221
[17] Klyshko D N 1986 *Physical Fundamentals of Quantum Electronics* (in Russian, Moscow: Mir)
[18] Yajima N and Oikawa M 1976 *Prog. Theor. Phys.* **56** 1719
[19] Sazonov S V 2012 *JETP Lett.* **96** 263
[20] Hirori H and Tanaka T 2016 *J. Phys. Soc. Japan* **85** 082001